Reversible and Quantum Circuits

Nabila Abdessaied • Rolf Drechsler

Reversible and Quantum Circuits

Optimization and Complexity Analysis

 Springer

Nabila Abdessaied
AG Rechnerarchitektur/
    Cyber-Physical Systems
University of Bremen/DFKI
Bremen, Germany

Rolf Drechsler
AG Rechnerarchitektur/
    Cyber-Physical Systems
University of Bremen
Bremen, Germany

ISBN 978-3-319-81158-1   ISBN 978-3-319-31937-7   (eBook)
DOI 10.1007/978-3-319-31937-7

Printed on acid-free paper

This springer imprint is published by Springer Nature
The registered company is Springer International Publishing AG Switzerland

Preface

Due to the increasing miniaturization according to Moore's law, classical circuit design will reach soon its limits. Reversible computation is an interesting alternative, since it provides a fundamental approach to the design of low-power circuits, a very critical aspect already in today's circuit design. Beyond that aspect, reversible circuits can be used as a starting point for synthesizing quantum circuits. Quantum computers can efficiently solve problems such as factorization, while the best known classical algorithms have sub-exponential complexity. In this emerging research field, first approaches to key design steps have been proposed, but they suffer from problems regarding complexity and scalability. Furthermore, most of the existing approaches are not optimal. Theoretical studies looking at the complexity of the underlying circuits are hard to perform, and so far only few results in this field are available.

In this context, this book introduces important new optimization approaches as well as complexity analysis that present various improvements compared to the state-of-the-art methods. The book has considered these two aspects on the different levels of the design flow of quantum circuits: the reversible level, the mapping level, and the quantum level. First, the book provides insight and guidance in all major aspects of optimizing the implementation of a reversible function as a circuit of reversible or quantum gates. Second, it gives an extensive overview of upper bounds on the number of gates needed in reversible and quantum circuits. New and tighter bounds are provided based on several synthesis approaches, mapping schemes, and gate libraries.

Bremen, Germany
January, 2016

Nabila Abdessaied
Rolf Drechsler

Acknowledgments

Foremost, we would like to thank all members of the Group of Computer Architecture at the University of Bremen and the Cyber-Physical Systems group of the German Research Center for Artificial Intelligence (DFKI) at Bremen that have supported us immensely by their precious advice and discussions.

We take this opportunity to express gratitude to our coauthors of the papers which were the basis of this book, including Mathias Soeken, Robert Wille, Matthew Amy, Gerhard W. Dueck, and Michael Kirkedal Thomsen, for their guidance, their successful collaboration, and their insightful suggestions.

Finally, special thanks go to D. Michael Miller, from the University of Victoria in Canada, for his valuable comments and proofreading.

Contents

1 Introduction 1
   1.1 Book Overview 4
       1.1.1 Optimization of Quantum Circuits 4
       1.1.2 Complexity Analysis 5
   1.2 Outline 6

2 Background 9
   2.1 Boolean Functions 9
   2.2 Boolean Function Decomposition 10
       2.2.1 Ashenhurst Decomposition 10
       2.2.2 Curtis Decomposition 11
       2.2.3 Bi-decomposition 12
       2.2.4 Multiplexer Decomposition 12
   2.3 Exclusive-OR Sum of Products 13
   2.4 Boolean Satisfiability and SAT Modulo Theory 15
   2.5 Reversible Logic 16
       2.5.1 Reversible Function 17
       2.5.2 Reversible Gates 20
       2.5.3 Reversible Circuits 22
   2.6 Quantum Computation 23
       2.6.1 Quantum Systems 24
       2.6.2 Quantum Libraries 26
       2.6.3 Quantum Circuits 30
   2.7 Cost Metrics for Reversible and Quantum Circuits 31
       2.7.1 Quantum Cost 31
       2.7.2 Number of Gates 33
       2.7.3 Number of Lines 34
       2.7.4 Depth 35
       2.7.5 Nearest Neighbor Cost 36

2.8   Decision Diagrams 36
        2.8.1   Binary Decision Diagrams 36
        2.8.2   Quantum Multiple-Valued Decision Diagrams................ 38

3   Optimizations and Complexity Analysis on the Reversible Level 45
3.1   Related Work.. 45
        3.1.1   Optimization Approaches of Reversible Circuits 46
        3.1.2   Complexity of Reversible Circuits 51
3.2   Exact Quantum Cost Optimization 52
        3.2.1   General Idea.. 52
        3.2.2   Encoding Using SMT 53
        3.2.3   Experimental Results 59
3.3   Heuristic Quantum Cost Optimization 65
        3.3.1   Simulated Annealing 66
        3.3.2   Rewriting Rules 67
        3.3.3   Algorithms 68
        3.3.4   Experimental Results 71
3.4   Complexity Analysis of Reversible Circuits 83
        3.4.1   Complexity of Single-Target Circuits........................ 83
        3.4.2   Complexity of MPMCT Circuits.............................. 84
        3.4.3   Upper Bounds for Single-Target Gates 85
        3.4.4   Upper Bounds for Reversible Circuits 87
3.5   Summary 89

4   Optimization and Complexity Analysis on the Mapping Level 91
4.1   Related Work.. 91
        4.1.1   Mapping Approaches... 92
        4.1.2   Complexity of NCT Circuits 100
4.2   Improving the Mapping of Single-Target Gates 100
        4.2.1   Motivation... 101
        4.2.2   Mapping of Single-Target Gates 101
        4.2.3   Experimental Evaluation 104
        4.2.4   Remarks and Observations 111
4.3   Improving the Mapping of MPMCT Gates to Clifford $+\ T$ Circuits . 112
        4.3.1   Clifford $+\ T$ Aware Reversible Circuit Mapping 112
        4.3.2   Proposed Mapping Approaches 113
        4.3.3   MPMCT Gates Mapping 114
        4.3.4   Experimental Results 122
4.4   Complexity Analysis of NCT Circuits 127
        4.4.1   Upper Bounds for MPMCT Gates 128
        4.4.2   Upper Bounds for Single-Target Gates 129
        4.4.3   Upper Bounds for NCT Circuits 138
4.5   Summary 140

5   Optimizations and Complexity Analysis on the Quantum Level 141
    5.1   Related Work 141
            5.1.1   Optimization of Quantum Circuits 141
            5.1.2   Complexity of Quantum Circuits 145
    5.2   Depth Optimization for NCV Circuits 145
            5.2.1   General Idea 147
            5.2.2   Optimization Approaches 148
            5.2.3   Experimental Results 153
    5.3   NCV-Cost Optimization 156
            5.3.1   Proposed Idea 157
            5.3.2   Application 158
            5.3.3   Experimental Results 159
    5.4   Complexity Analysis of Quantum Circuits 163
            5.4.1   Complexity of NCV Quantum Circuits 163
            5.4.2   Complexity of Clifford+T Quantum Circuits 169
    5.5   Summary 174

6   Conclusions 175

References 179

Acronyms

BDD	Binary Decision Diagram
CMOS	Complementary Metal Oxide Semiconductor
CNF	Conjunctive Normal Form
CNOT	Controlled NOT
DPLL	Davis Putnam Logemann Loveland
ESOP	Exclusive-OR Sum-of-Product
FPRM	Fixed Polarity Reed-Muller expression
KRO	Kronecker Expression
MCT	Multiple-Control Toffoli
MPMCT	Mixed-Polarity Multiple-Control Toffoli
NCT	NOT, CNOT, Toffoli
NCV	NOT, CNOT, V
NNC	Nearest Neighbor Cost
PPRM	Positive Polarity Reed-Muller Expression
PSDKRO	Pseudo-Kronecker Expression
QMDD	Quantum Multiple-Valued Decision Diagram
Qubit	Quantum Bit
SAT	Boolean Satisfiability
SMT	Satisfiability Modulo Theory
ST	Single Target

List of Figures

Fig. 1.1   Design flow for quantum circuits inspired from [107]   2

Fig. 1.2   Mapping model   6

Fig. 2.1   Boolean functions. (**a**) ($\vee$) Function. (**b**) 1-bit full adder function .   10

Fig. 2.2   Ashenhurst decomposition   11

Fig. 2.3   Curtis decomposition   11

Fig. 2.4   Bi-decomposition   12

Fig. 2.5   Multiplexer decomposition   13

Fig. 2.6   DPLL algorithm in modern SAT solvers   16

Fig. 2.7   Examples of Boolean functions. (**a**) f_1: Irreversible
function. (**b**) f_2: Reversible function   17

Fig. 2.8   Embedding of the 1-bit full adder function. (**a**) 1-bit
full adder function. (**b**) Garbage outputs. (**c**) Constant
input   19

Fig. 2.9   Single-target gate representation   20

Fig. 2.10   Toffoli library. (**a**) MPMCT gate. (**b**) MCT gate.
(**c**) Toffoli gate. (**d**) CNOT gate. (**e**) NOT gate   21

Fig. 2.11   ST gate with a controlling function ($\vee$) given by
Fig. 2.1b. (**a**) ST gate. (**b**) MPMCT circuit. (**c**) MCT
circuit gate   22

Fig. 2.12   Reversible circuit structure inspired from [107].....................   23

Fig. 2.13   Reversible circuits for 1-bit full adder as specified in
Fig. 2.8. (**a**) MPMCT library based circuit. (**b**) MCT
library based circuit...   23

Fig. 2.14   NCV quantum mapping of different Toffoli gates. (**a**)
Toffoli with two positive controls. (**b**) Toffoli with
a negative and a positive control. (**c**) Toffoli with a
positive and a negative control. (**d**) Toffoli with a
negative controls   28

Fig. 2.15   Clifford + T quantum mapping of different Toffoli
             gates. (**a**) Toffoli with two positive controls. (**b**) Toffoli
             with a negative and a positive control. (**c**) Toffoli with
             a positive and a negative control. (**d**) Toffoli with a
             negative controls 29
Fig. 2.16   Mapping of the 1-bit full adder reversible circuit
             in Fig. 2.13b. (**a**) NCV based quantum circuit. (**b**)
             Clifford + T based quantum circuit................................. 30
Fig. 2.17   Examples of entangled and SCQC quantum circuits.
             (**a**) Entangled circuits. (**b**) SCQC circuits 31
Fig. 2.18   Quantum circuits realizing a 1-bit full adder. (**a**)
             Optimal NCV circuit. (**b**) Optimal Clifford + T circuit 32
Fig. 2.19   MPMCT based reversible circuit..................................... 33
Fig. 2.20   Reversible circuits realizing a 1-bit full adder. (**a**) First
             realization. (**b**) Second realization. (**c**) Optimal circuit
             regarding the gate-count.. 34
Fig. 2.21   Quantum depth for a 1-bit full adder circuit......................... 35
Fig. 2.22   NCV quantum circuits realizing the Toffoli gate given
             in Fig. 2.10c. (**a**) First realization. (**b**) Second realization 36
Fig. 2.23   Binary decision diagrams. (**a**) Shannon decomposition.
             (**b**) BDD for a 1-bit full adder..................................... 38
Fig. 2.24   Unitary matrix representations. (**a**) Quantum circuit
             based on the NCV library. (**b**) Transformation matrix.
             (**c**) QMDD.. 39
Fig. 2.25   1-bit full adder reversible function and its different
             representations. (**a**) Truth table. (**b**) Permutation
             matrix. (**c**) QMDD 42

Fig. 3.1   Moving rules. (**a**) Classical moving rule. (**b**) Extended
             moving rule-based on LLP 46
Fig. 3.2   Examples of MPMCT rules. (**a**) Rule 1. (**b**) Rule 2. (**c**)
             Rule 3. (**d**) Rule 4.. 47
Fig. 3.3   Rule-based optimization... 48
Fig. 3.4   Templates with 2 or 3 inputs from [79]. (**a**) Template 1.
             (**b**) Template 2. (**c**) Template 3. (**d**) Template 4 48
Fig. 3.5   Template disposition. (**a**) Original template. (**b**)
             Duplication. (**c**) Removal. (**d**) Replacement. (**e**) Rotation 49
Fig. 3.6   Optimizing a reversible circuit based on template matching 50
Fig. 3.7   Optimizing a reversible circuit with one additional line
             taken from [152]. (**a**) Original circuit. (**b**) With one
             additional line 51
Fig. 3.8   Encoding the search method as an SMT instance 53
Fig. 3.9   Reversible circuit and template. (**a**) Circuit. (**b**) Template.......... 54
Fig. 3.10   Circuit encoding 55
Fig. 3.11   Template encoding... 56

Fig. 3.12   Moving rule graph for circuit in Fig. 3.9a   57
Fig. 3.13   Gate-count evaluation   63
Fig. 3.14   Time evaluation. (**a**) Run time of the heuristic and
            exact template matching algorithms. (**b**) Correlation of
            the ETM runtime and the circuit lines   64
Fig. 3.15   Quantum cost evaluation of different synthesis
            approaches. (**a**) NCV-cost. (**b**) T-depth   65
Fig. 3.16   Greedy heuristics via simulated annealing algorithm   67
Fig. 3.17   Rewriting rules. (**a**) Rule 1. (**b**) Rule 2. (**c**) Rule 3. (**d**)
            Rule 4   67
Fig. 3.18   Circuit optimization using different moving rules. (**a**)
            Original realization. (**b**) Classical moving rule. (**c**)
            Extended moving rule. (**d**) Rewriting rules..........................   68
Fig. 3.19   Applied reductions. (**a**) Rule 1. (**b**) Rule 2. (**c**) Rule 3.
            (**d**) Rule 4. (**e**) Rule 5   69
Fig. 3.20   Evaluation of optimization approaches regarding the
            quantum cost. (**a**) NCV-cost evaluation. (**b**) T-depth evaluation ...   72
Fig. 3.21   Time evaluation   80
Fig. 3.22   Quantum cost of optimized circuits obtained from
            different synthesis approaches. (**a**) NCV-cost
            evaluation. (**b**) T-depth evaluation   81
Fig. 3.23   Quantum cost resulting from SA and ETM. (**a**)
            NCV-cost evaluation. (**b**) T-depth evaluation   82
Fig. 3.24   Synthesis based on Young subgroups................................   85
Fig. 3.25   Single-target gate decomposition. (**a**) MPMCT case.
            (**b**) MCT case   86
Fig. 4.1   Quantum mapping of a Toffoli gate. (**a**) Toffoli gate.
            (**b**) NCV circuit. (**c**) Clifford + T circuit   92
Fig. 4.2   Mapping an MCT gate using the Barenco
            et al. (Lemma 7.2). (**a**) 4-control gate. (**b**) Barenco
            et al. (Lemma 7.2)   92
Fig. 4.3   NCV-cost optimized Barenco et al. (Lemma 7.2)
            mapping. (**a**) 4-control gate. (**b**) Barenco
            et al. (Lemma 7.2). (**c**) Toffoli gates replacement. (**d**)
            Optimized mapping regarding the NCV-cost........................   94
Fig. 4.4   Mapping an MCT gate using the Barenco
            et al. (Lemma 7.3). (**a**) 7-control gate. (**b**) Barenco
            et al. (Lemma 7.3). (**c**) Final mapping based on
            Barenco et al. mapping   95
Fig. 4.5   Mapping an MCT gate using the Nielsen and Chuang
            mapping. (**a**) 7-control gate. (**b**) Nielsen and Chuang
            mapping. (**c**) Final mapping based on Nielsen and
            Chuang mapping...   96

Fig. 4.6   Mapping an MPMCT gate using the Miller
            et al. mapping. (**a**) 7-control gate. (**b**) Miller
            et al. mapping. (**c**) Final mapping based on Miller et al. mapping .   98
Fig. 4.7   Mapping of multiple targets gate. (**a**) MCT gates with
            common controls. (**b**) Multiple target gate. (**c**) New mapping......   99
Fig. 4.8   Mapping flows for single-target gate based circuits. (**a**)
            Mapping flow for a single-target gate. (**b**) Proposed
            mapping flow for a single-target gate   102
Fig. 4.9   Different types of decomposition. (**a**)
            Ashenhurst-Curtis. (**b**) MUX. (**c**) AND ($\wedge$).
            (**d**) OR ($\vee$). (**e**) XOR ($\oplus$). (**f**) XNOR ($\leftrightarrow$)   103
Fig. 4.10   Example of mapping a single-target gate. (**a**) Mapping
            example using the classical mapping flow. (**b**) Mapping
            example using the new mapping flow..............................   106
Fig. 4.11   Quantum cost resulting from classical and proposed
            approaches. (**a**) NCV-cost evaluation. (**b**) T-depth
            evaluation   108
Fig. 4.12   Time evaluation   111
Fig. 4.13   Optimized B2 mapping with respect to T-depth. (**a**)
            5-control gate. (**b**) B2 mapping (Lemma 7.2). (**c**)
            Toffoli gates replacement. (**d**) Optimized B2 mapping   116
Fig. 4.14   Optimization of B2 mapping with respect to T-depth.
            (**a**) Toffoli gates replacement. (**b**) Swoping the target
            and a control of the Z gates. (**c**) Z gates replacement.
            (**d**) Redudant gates reduction.......................................   118
Fig. 4.15   Clifford $+ T$ implementations for mixed-polarity $i\omega Z$
            gates. (**a**) iZ gate with two positive controls. (**b**) iZ gate
            with a negative and a positive control. (**c**) iZ gate with
            negative controls   119
Fig. 4.16 T-depth of the 5-control MCT gate using B2 mapping.
            (**a**) iZ gate replacement. (**b**) Redundant gates reduction.
            (**c**) Final mapping result   120
Fig. 4.17 T-depth of mapped MPMCT gates and circuits. (**a**)
            T-depth of MPMCT gates using B1. (**b**) T-depth of
            MPMCT gates using B2 mapping. (**c**) T-depth of
            MPMCT gates using one ancilla mappings.........................   123
Fig. 4.18 T-depth of circuits obtained from different synthesis approaches..   127
Fig. 4.19 T-depth of mapped circuit using one ancilla mapping algorithms .   127
Fig. 4.20   ST gate decomposition based on MCT gates. (**a**) ST
            gate decomposition. (**b**) MCT circuit   136

Fig. 4.21   ST gate decomposition based on MPMCT gates. (**a**) ST
            gate decomposition. (**b**) MPMCT circuit 137
Fig. 5.1   Quantum templates with 2 or 3 lines. (**a**) Template
            1. (**b**) Template 2. (**c**) Template 3. (**d**) Template 4.
            (**e**) Template 5. (**f**) Template 6. (**g**) Template 7. (**h**)
            Template 8 142
Fig. 5.2   Reduction rules for the NCV circuits. (**a**) Rule 1. (**b**)
            Rule 2. (**c**) Rule 3 142
Fig. 5.3   Simplification of an NCV circuit for the 1-bit full
            adder taken from [81]. (**a**) Original circuit. (**b**) Mapped
            circuit. (**c**) Optimized circuit 142
Fig. 5.4   LLP moving rule 143
Fig. 5.5   DDMF moving rule 143
Fig. 5.6   An optimal T-depth implementation of the 1-bit full adder 144
Fig. 5.7   A T-depth 1 implementation of a 1-bit full adder 145
Fig. 5.8   Quantum depth. (**a**) Reversible circuit. (**b**) Quantum
            circuit. (**c**) Depth computing 147
Fig. 5.9   Depth reduction using an ancilla. (**a**) Initial circuit. (**b**)
            Circuit with reduced depth 148
Fig. 5.10   Proposed depth-aware mapping for each single gate.
            (**a**) Toffoli gate. (**b**) Original mapping. (**c**) Proposed
            mapping. (**d**) Peres gate. (**e**) Original mapping. (**f**)
            Proposed mapping. (**g**) Inverse Peres gate. (**h**) Original
            mapping. (**i**) Proposed mapping. (**j**) αP gate. (**k**)
            Original mapping 149
Fig. 5.11   Application of the local scheme to the circuit from Fig. 5.8a....... 149
Fig. 5.12   Application of the local scheme to concurrent Toffoli
            gates. (**a**) Original circuit. (**b**) Original mapping. (**c**)
            Proposed mapping 151
Fig. 5.13   Consideration of the whole circuit. (**a**) Original circuit.
            (**b**) Resulting circuit 152
Fig. 5.14   Application of global scheme to the circuit from Fig. 5.8a 152
Fig. 5.15   Optimizing quantum circuits based on equivalence checking 157
Fig. 5.16   Gate rearrangement based on QMDD equivalence checking 158
Fig. 5.17   Optimization using different filters 159
Fig. 5.18   Time evaluation of the used filters 163

Fig. 6.1   Book contributions... 176

List of Tables

Table 1.1   Overview of the presented bounds................................... 7

Table 2.1   Number of reversible gates using n lines 21
Table 2.2   Gate definitions 27
Table 2.3   Possible qubit values for NCV gates 28
Table 2.4   Number of quantum gates using n lines 29
Table 2.5   Quantum cost for MPMCT gates with c controls 32

Table 3.1   Experimental results for heuristic and SMT based
             template matching 61
Table 3.2   Experimental results for line labeling and greedy approaches 74
Table 3.3   Experimental results for greedy and simulated
             annealing approaches............................... 76
Table 3.4   Quantum cost variation.............................. 78
Table 3.5   Summary of upper bounds for realizing a single-target gate 87
Table 3.6   Summary of upper bounds for representing a reversible function . 88
Table 3.7   Upper bounds on number of gates in a reversible circuit 88

Table 4.1   Number of NCT gates and quantum cost of a c-control
             MPMCT gate 98
Table 4.2   Experimental results for mapping ST circuits..................... 109
Table 4.3 T-depth for a c-control MPMCT gate 114
Table 4.4   Experimental results for mapping MPMCT circuits 125
Table 4.5   Number of NCT gates for an MPMCT gate with c controls 128

Table 5.1   Experimental results for the depth optimization approaches 154
Table 5.2   Experimental results for NCV-cost optimization approaches 161
Table 5.3   NCV-cost for an MPMCT gate with c controls 164
Table 5.4   Updated T-depth for an MPMCT gate with c controls 169

Table 6.1   Complexity of reversible and quantum circuits 177
Table 6.2   Summary for gate and circuit complexity 177

Chapter 1
Introduction

During the last decades, the number of transistors in integrated circuits has been doubled approximately every 18 months (also known as "Moore's Law"). Even if this procedure may continue in the upcoming years, the conventional computer hardware technologies are going to reach their limits in the near future. In particular, the power dissipation is going to become crucial. While nowadays the power dissipation is particularly caused, e.g., by nonideal behavior of transistors and materials, a more fundamental limit (namely "Landauers Barrier" [69]) will be approached. Landauer stated that each time information is lost during computation, the energy is dissipated. Extrapolating the recent achievements in the development of classical CMOS technologies, this amount of power dissipation will halt further miniaturization in the near future [157].

As a consequence, researchers intensely studied alternative technologies and in this context, reversible circuits are promising. In fact, they allow bijective operations only, thereby, they do not lose information during the computation and, thus, can avoid power dissipation caused by information loss [18, 98]. Reversible computation [38] has proven itself as a very promising research area, especially for applications to emerging technologies. This is confirmed by its results in emerging applications such as quantum computation [66, 98, 125, 141] which has shown promising results for solving certain problems exponentially faster than any known classical algorithm by exploiting quantum mechanical effects [42, 43]. In contrast to Boolean logic, quantum bits (qubits) represent not only the classical 0 and 1 states but also any complex combination or *superposition* of both, leading to a significant speed-up in computing. It has been shown that, for example, using a quantum circuit makes it possible to solve the factorization problem [125] in polynomial time using the factorization algorithm by Shor [125, 141], while for classical circuits only sub-exponential methods are known. Furthermore, the Deutsch-Jozsa algorithm [41] as well as the Grover search [56] algorithm are famous examples showing the power of

© Springer International Publishing Switzerland 2016
N. Abdessaied, R. Drechsler, *Reversible and Quantum Circuits*,
DOI 10.1007/978-3-319-31937-7_1

quantum computing. The design and fabrication of these machines have progressed rapidly in the past decade, with many research groups now routinely fabricating and operating small quantum computers in physical systems [28, 97].

Quantum computing does not only provide challenges for physicists but also offers a variety of challenging and interesting problems to the field of computer science. Large parts of quantum computers perform classical computations which can be described in terms of classical Boolean functions instead of arbitrary unitary operations as are used for general quantum computing. However, since all quantum transformations need to be reversible, the classical computations need to be described in terms of reversible functions [98] to be used in the model. But since reversible logic is subject to certain characteristics and restrictions, the design methodology of circuits and systems following the reversible computation paradigm significantly differs from the established (classical) design flow. Many concepts and methods developed for classical hardware design in the last decades have to be redeveloped in order to support these new technologies. Accordingly, significant progress has been made in the design of reversible and quantum circuits in the last years. The synthesis of a quantum circuit from such a Boolean function is typically conducted by a three-stage (see Fig. 1.1) procedure: (1) a reversible circuit realizing the Boolean component is generated for which existing reversible logic synthesis algorithms such as [46, 85, 135] are used, (2) mapping techniques are applied to

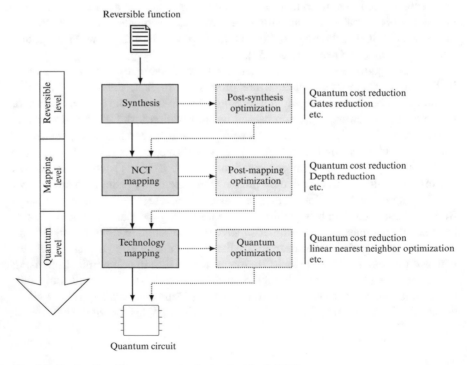

Fig. 1.1 Design flow for quantum circuits inspired from [107]

transform the reversible circuit into a functionally equivalent reversible circuit that consist of specific gates [15, 91, 98], and (3) the technology mapping of the resulting circuits to quantum circuits that consist only of quantum gates from a determined library [15, 92, 133]. Each of these steps present a major level in the design flow as shown in Fig. 1.1: reversible level, mapping level, and quantum level. Within each level, an optional process is applied to improve the obtained result.

The flow starts with the *reversible level* where the synthesis of reversible circuit takes place. Synthesis algorithms describe techniques to find a reversible circuit that realizes a given reversible function. The most common gate libraries to realize reversible circuits consist of *mixed-polarity multiple-controlled Toffoli* gates (MPMCT) or *single-target* gates (ST). Several approaches have been proposed [46, 85, 132, 135, 145] to synthesize a given function. Typically, the majority of these approaches do not guarantee optimal realizations, in fact, the algorithms that do guarantee an optimal solution (e.g., [146]) are only applicable to small circuits having four to six lines. As a result, in order to optimize synthesis results with respect to some target technology, several post-synthesis optimization approaches [88, 128] are used to reduce the circuits with respect to a given cost metric. Among the cost metrics we find the number of gates and the number of lines. Other important metrics are the number of quantum gates and the gate delay (depth).

MPMCT gates and ST gates offer a convenient representation to model the functionality of a reversible circuit but they are still too abstract to be used as quantum operations. Many aspects, particularly those considering fault tolerance and error correction properties, cannot be effectively considered on that abstraction level. Therefore, reversible circuits are mapped to quantum circuits that only consist of gates from a particular gate library. For this purpose, the reversible circuit is mapped to a functionally equivalent Toffoli circuit in the *mapping level*. In other words, each ST gate and Toffoli gate in the circuit with more than two control lines is mapped into a circuit of Toffoli gates with at most two control lines (NCT). The MPMCT gates mapping algorithms were introduced in [15, 91, 98]. Afterwards, the mapped circuit is optimized with respect to technology dependent cost metrics [75, 150] in the post-mapping step.

In the *quantum level*, the technology mapping step maps reversible circuits to quantum circuits that only consist of gates of a gate library suitable for the target technology. For this purpose, each Toffoli gate of the reversible circuit is replaced, with respect to a particular gate library, with an optimal circuit of functionally equivalent quantum gates [10, 15]. Quantum libraries consist of a few quantum gates that typically act on at most 2 qubits: one of the currently prominent libraries is the Clifford$+T$ [98] library and the NCV gate [15] library. After mapping the reversible circuit to its equivalent quantum circuit, the obtained quantum circuit is optimized by applying algorithms that aim at obtaining cheaper realizations.

While this layered approach allows convenient reuse of circuits between applications, the resulting circuits are typically far from optimal, particularly because the structure of the high-level circuit may allow gates to be omitted [76, 107, 150]. As a result, researchers have proposed cost-aware optimization algorithms [75, 88, 91]

from reversible circuits into intermediate or low-level quantum gate libraries. Such optimizations avoid the overhead of applying primitive gate optimization and achieve better results [12, 86].

Following these methods, this book presents solutions to the major obstacles facing the optimization of quantum circuits (e.g., gate movement, cost reduction, applied heuristics) in the different levels of its design flow shown in Fig. 1.1. Several cost-aware optimization techniques for quantum circuit are introduced on the reversible level [1, 6], the mapping level [5, 8], and the quantum level [2, 3, 9]. Since the nonoptimality of quantum circuits derives mainly from the exponential complexity of reversible circuits [108], other work of this book investigates the required number of gates to implement reversible functions and, hence, presents tighter upper bounds [4, 7, 134] for the number of gates needed in a reversible or quantum circuit.

1.1 Book Overview

This book focuses on two aspects. First, the optimization of reversible and quantum circuits. Second, an extensive study of the complexity of reversible gates and reversible circuits as well as quantum circuits is presented.

1.1.1 *Optimization of Quantum Circuits*

At the *reversible level*, two new different optimization approaches which target the reduction of the quantum cost are presented. So far, the existing optimization approaches at this level are usually heuristic and employ restricted moving rules to find local optimizations. Therefore, a new optimization algorithm [6] that handles positive and negative controls and allows more freedom to rearrange gates into a circuit compared to the existing moving rules is given. Furthermore, since existing optimizations cannot guarantee optimality for small circuits, a new approach is proposed [1] that exploits Boolean satisfiability techniques allowing an exhaustive yet efficient search of any possible reduction and hence ensures optimal results. For larger realizations, the application of simulated annealing is considered to find further possible reductions that could not be found using existing approaches.

At the *mapping level*, a new mapping algorithm [5] for decomposing single-target gates is described. Since each ST gate contains a Boolean control function, the given method attempts to find smaller ST gates based on a functional decomposition of the control function. It consists of breaking a large ST gate into smaller ones using additional lines. This technique leads to circuits with cheaper quantum cost. Furthermore, a new optimized mapping methodology [9] to decompose reversible circuits consisting of MPMCT gates into quantum circuits is introduced.

At the *quantum level*, new algorithms [3, 9] that aims at reducing the quantum cost for a given NCV or Clifford $+$ T based circuit are described. Furthermore, an optimization technique for reducing the depth of NCV based circuits using additional lines [2] is proposed.

1.1.2 Complexity Analysis

To compare the efficiency of different synthesis approaches, the resulting circuits are evaluated in terms of cost metrics which depend on the targeted technology. Among these metrics the number of gates and quantum cost present the major criteria in the evaluation process. As a second direction, throughout this book, a detailed overview studying the complexity in terms of the number of gates or the quantum cost for a given circuit in different levels is given. Tighter upper bounds for the number of gates that are required to implement a reversible function in a reversible or quantum circuit are determined. This is important as an approach to understand the complexity of reversible circuits, but also to give an overall quality measure of the different reversible synthesis methods. Previous research has investigated this topic based on specific synthesis algorithms and using a specific gate library [76, 108]. This book presents extensions of the upper bounds to more gate libraries. As reversible gate libraries, The complexity analysis covered reversible circuits based on reversible gate libraries (ST gates, MPMCT gates, and NCT gates) and quantum circuits based on quantum gate libraries (the semi-classical NCV library [15] and the Clifford $+$ T library).

This book provides upper bounds for a reversible or quantum gate library on each level of the design flow. Figure 1.2 illustrates the layered mapping approach from which the upper bounds have been derived. For instance, in the reversible level, new upper bounds on the number of ST gates required to synthesize any reversible function [4] are presented. Along with the number of gates required for the realization of a reversible function, in the case of MPMCT gates it is further investigated how many of such gates are required to realize one single-target gate. After mapping these circuits into circuits that are solely composed of NCT gates based on three different mapping schemes, the complexity of reversible gates and circuits at this level [134] is derived. Finally, after mapping NCT circuits into quantum circuits with respect to a quantum gate library, new upper bounds [7] on the quantum cost to implement ST gates, MPMCT gates, and reversible circuits using the NCV library as well as the Clifford $+$ T library are derived. The book provides several new and tighter bounds that are calculated using several synthesis approaches and mapping schemes.

Table 1.1 gives an overview of all the considered configurations. The first column lists all the gate libraries while the remaining columns describe the function class for which bounds are given.

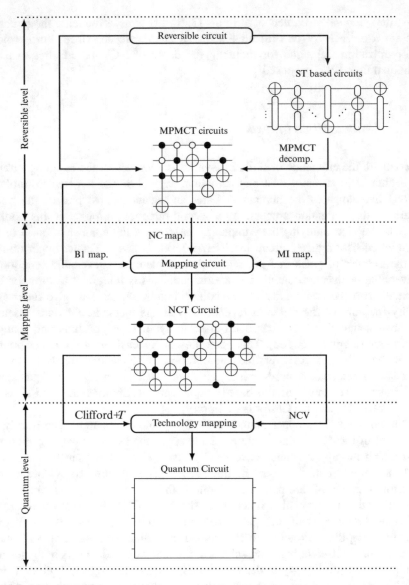

Fig. 1.2 Mapping model

1.2   Outline

This book consists of three main chapters which cover the optimization and complexity analysis of quantum circuits in the three major level of their design flow: (1) the reversible level, (2) the mapping level, and (3) the quantum level. These

Table 1.1 Overview of the presented bounds

Gate library	One MPMCT gate	One ST gate	Reversible circuit
ST gates			Section 3.4.1
MPMCT gates		Section 3.4.3	Section 3.4.4
NCT gates	Section 4.4.1	Section 4.4.2	Section 4.4.3
NCV gates	Section 5.4.1.1	Section 5.4.1.2	Section 5.4.1.3
Clifford $+$ T gates	Section 4.3.3	Section 5.4.2.2	Section 5.4.2.3

chapters are preceded by a background chapter and summarized by conclusions. The following outlines the main aspects of this book chapter by chapter.

Chapter 2 provides the required background to keep this book self-contained. First, Boolean functions and decomposition are introduced. Since several of the algorithms in the book are based on formal representations of functions, Boolean proof techniques such as SAT solvers are briefly reviewed. Reversible logic, quantum computation, and cost metrics for reversible and quantum circuits are introduced next. The chapter concludes with a review of decision diagrams with sufficient examples and references.

Chapter 3 considers optimization and complexity on the reversible level. The chapter starts with a discussion of previous work. This allows the reader to easily identify the contribution of this book. Various optimization techniques, such as exact algorithms, rule-based approaches, or simulated annealing, are presented. All proposed algorithms have been implemented and evaluated on benchmark functions. This allows to understand the effect of the individual steps very well. The second part of the chapter provides a very detailed analysis of the complexity of reversible circuits. Based on the complexity analysis existing upper bounds have been improved and extended to more gate libraries.

Chapter 4 considers optimization and complexity analysis at the mapping level where individual reversible gates are mapped to sequences of less complex gates. The chapter starts with a discussion of previous work. Several examples are given that allow for an easy understanding of the complex topic. Then, two improved mapping algorithms for ST gates and MPMCT are discussed and experimentally evaluated. Finally, several upper bounds are proven on the complexity of the resulting circuits in this level.

Chapter 5 presents new optimization methods to improve quantum circuits in different quantum libraries. Furthermore, it gives new complexity analysis for quantum circuits. Here, the reduction of the depth of the circuit is considered as well as the quantum cost. Again, all proposed algorithms have been implemented and evaluated on benchmark functions. This chapter concludes with complexity analyses of quantum circuits based on two different gate libraries.

Chapter 6 summarizes the results of the book and provides directions for possible future work.

Chapter 2
Background

To keep the book self-contained, this chapter briefly introduces the basics on reversible logic, quantum circuits, and other required concepts. The chapter consists of seven parts: the first section introduces the basic definitions and notations of Boolean functions while the second section reviews existing Boolean decomposition. The third and the fourth sections give an overview of exclusive sum of products and Boolean satisfiability, respectively. The fifth section provides a summary on the principles of reversible logic required later in this book. Similarly, the sixth section gives an introduction on quantum computation and quantum circuits. The later section details the different metrics used as a quality-measure of reversible and quantum realizations. Finally, decision diagrams are reviewed.

2.1 Boolean Functions

A *Boolean function* is defined as follows.

Definition 2.1 (Boolean Function). Let $\mathbb{B} \stackrel{\text{def}}{=} \{0, 1\}$ denote the *Boolean values*. Then we refer to

$$\mathscr{B}_{n,m} \stackrel{\text{def}}{=} \{f \mid f \colon \mathbb{B}^n \to \mathbb{B}^m\}$$

as the set of all *Boolean functions* with n inputs and m outputs. There are $2^{m 2^n}$ such Boolean functions. We write $\mathscr{B}_n \stackrel{\text{def}}{=} \mathscr{B}_{n,1}$ to denote the set of Boolean 1-output functions.

Assume that each multiple-output function $f \in \mathscr{B}_{n,m}$ is represented as a tuple $f = (f_1, \ldots, f_m)$ where $f_i \in \mathscr{B}_n$ for each $i \in \{1, \ldots, m\}$ and hence $f(X) = (f_1(X), \ldots, f_m(X))$ for each $X \in \mathbb{B}^n$. The functions $f_i(X)$ are called *primary outputs*.

© Springer International Publishing Switzerland 2016
N. Abdessaied, R. Drechsler, *Reversible and Quantum Circuits*,
DOI 10.1007/978-3-319-31937-7_2

Fig. 2.1 Boolean functions.
(**a**) (∨) Function. (**b**) 1-bit
full adder function

a

x_1	x_2	y
0	0	0
0	1	1
1	0	1
1	1	1

b

x_1	x_2	C_{in}	C_{out}	S
0	0	0	0	0
0	0	1	0	1
0	1	0	0	1
0	1	1	1	0
1	0	0	0	1
1	0	1	1	0
1	1	0	1	0
1	1	1	1	1

The behavior of a Boolean function can be specified by enumerating the outputs respond to the inputs values. This input/output relationship is commonly enumerated in a tabular form, called a *truth table* in which all permutations of the inputs are listed on the left, and the outputs of the function are listed on the right.

Example 2.1. Figure 2.1a shows the truth table of a function with 2 inputs and 1 output. This function represents the OR function (∨). Figure 2.1b presents the truth table of a multi-output function with 3 inputs and 2 outputs. This specification presents the behavior of the 1-bit full adder.

2.2 Boolean Function Decomposition

Boolean function decomposition describes the problem of expressing a Boolean function of n variables as a functionally equivalent composition of simpler Boolean functions of less than n variables. Several types of Boolean function decomposition that have been proposed in the last decades are outlined in the following.

2.2.1 Ashenhurst Decomposition

When a function $f \in \mathcal{B}_n$ can be decomposed (see Fig. 2.2) into

$$f(X) = h(g(X_1, X_2), X_2, X_3)$$

Fig. 2.2 Ashenhurst
decomposition

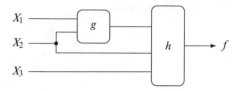

Fig. 2.3 Curtis
decomposition

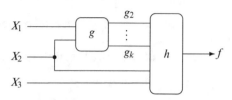

with $g \in \mathscr{B}_{|X_1|+|X_2|}$, $h \in \mathscr{B}_{|X_2|+|X_3|+1}$, $X = X_1 \cup X_2 \cup X_3$, and $X_1 \cap X_2 = X_1 \cap X_3 = X_2 \cap X_3 = \emptyset$, then it is referred to as *Ashenhurst decomposition*. If $X_2 = \emptyset$, then the decomposition is called *disjoint*, otherwise it is called a *non-disjoint* decomposition. The set X_1 is called *bound set* and the set X_3 is called *free set*.

Example 2.2 ([155]). Consider the Boolean function

$$f = x_1\bar{x}_2 \vee x_1 x_3 \vee x_1 \bar{x}_4 \vee \bar{x}_2 \bar{x}_3 x_4$$

f can be decomposed using the Ashenhurst decomposition. One possible disjoint decomposition of f is as follows.

$$f(x_1, x_2, x_3, x_4) = h(x_1, x_2, g)$$
$$g(x_3, x_4) = \bar{x}_3 x_4$$
$$h(x_1, x_2, g) = \bar{x}_2 g \vee x_1 \bar{g}$$

2.2.2 Curtis Decomposition

A generalization of the Ashenhurst decomposition was suggested by Curtis [31] where g can have multiple outputs (see Fig. 2.3), i.e.,

$$f(X) = h(g_1(X_{1_1}), g_2(X_{1_2}), \dots, g_k(X_{k}), X_2, X_3)$$

with $g_i \in \mathscr{B}_{|X_{j_i}|}$ (where $i = 1 \cdots k$ and $j = 1, 2$), $h \in \mathscr{B}_{|X_2|+|X_3|+k}$, $X = X_1 \cup X_2 \cup X_3$, and $X_1 \cap X_2 \cap X_3 = \emptyset$.

Example 2.3 ([155]). Consider the Boolean function

$$f = \bar{x}_1 x_2 x_3 x_4 \bar{x}_5 \vee \bar{x}_1 \bar{x}_4 (\bar{x}_2 \vee \bar{x}_3) \vee x_1 x_2 x_3 x_4 \vee x_1 x_4 (\bar{x}_2 \vee \bar{x}_3) \vee x_1 x_5 \vee \bar{x}_4 x_5.$$

Fig. 2.4 Bi-decomposition

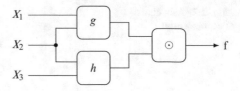

One possible disjoint Curtis decomposition associated with f is as follows.

$$f(x_1, x_2, x_4, x_5) = h(g_1, g_2, x_4, x_5)$$

$$g_1(x_1, x_2, x_3) = x_1$$

$$g_2(x_1, x_2, x_3) = x_1\bar{x}_2 \vee x_2 x_3$$

$$h(g_1, g_2, x_4, x_5) = \bar{g}_1\bar{g}_2\bar{x}_4 \vee \bar{g}_1 g_2(x_4 \oplus x_5) \vee g_1\bar{g}_2 x_4 \vee g_1 g_2\bar{x}_4$$

2.2.3 Bi-decomposition

When a Boolean function can be decomposed into two sub-functions [117] (see Fig. 2.4) $f(X) = g(X_1) \odot h(X_2)$, with $g \in \mathscr{B}_{|X_1|}$, $h \in \mathscr{B}_{|X_2|}$, $X = X_1 \cup X_2$, and $X_1 \cap X_2 = \emptyset$, then the decomposition is called *bi-decomposition*. The '$\odot$' can be any binary Boolean operation (typically $\vee$, $\wedge$, $\oplus$, or $\otimes$). This decomposition is also known as the *simple decomposition*.

Example 2.4 ([155]). Consider the Boolean function

$$f = \bar{x}_1 x_2 \bar{x}_3 \vee x_1 x_2 \bar{x}_3 \vee x_1 \bar{x}_2 x_3.$$

f can be decomposed into the following non-disjoint bi-decomposition:

$$f(x_1, x_2, x_3) = g(x_1, x_2, x_3) \wedge h(x_2, x_3)$$

$$g(x_1, x_2, x_3) = x_1 x_3 \vee x_2$$

$$h(x_2, x_3) = \bar{x}_2 \vee \bar{x}_3$$

2.2.4 Multiplexer Decomposition

If a Boolean function can be decomposed into three sub-functions (see Fig. 2.5)

$$f(X) = l(X_4) \wedge g(X_1, X_3) \oplus \bar{l}(X_4) \wedge h(X_2, X_3)$$

Fig. 2.5 Multiplexer
decomposition

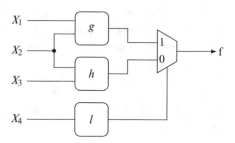

with $g \in \mathscr{B}_{|X_1|+|X_2|}$, $h \in \mathscr{B}_{|X_2|+|X_3|}$, $l \in \mathscr{B}_{|X_4|}$, $X = X_1 \cup X_2 \cup X_3 \cup X_4$, and $X_1 \cap X_2 = X_1 \cap X_3 = X_2 \cap X_3 = \emptyset$, then it is referred to as *multiplexer decomposition*.

Example 2.5 ([155]). Consider the Boolean function $f(x_1, x_2, x_3) = \bar{x}_1 x_2 x_3 \vee x_1 \bar{x}_2 x_3 \vee x_1 x_2 \bar{x}_3$ In the following, a possible multiplexer decomposition for f is given:

$$f(x_1, x_2, x_3) = l(x_1)g(x_2, x_3) \oplus \bar{l}(x_1)h(x_2, x_3)$$
$$h(x_2, x_3) = x_2 x_3$$
$$g(x_2, x_3) = x_2 \oplus x_3$$
$$l(x_1) = x_1$$

2.3   Exclusive-OR Sum of Products

Exclusive-OR sum-of-products (ESOPs, [113])

$$f = \bigoplus_{i=1}^{k} x_{i_1}^{p_{i_1}} \wedge \cdots \wedge x_{i_{l_i}}^{p_{i_{l_i}}} \tag{2.1}$$

are two-level descriptions for Boolean functions in which a function is composed of k *product terms* that are combined using the exclusive-OR (EXOR, $\oplus$) operation. A product term is the conjunction of l_i literals where a *literal* is either a propositional variable $x^1 = x$ or its negation $x^0 = \bar{x}$. ESOPs are the most general form of two-level AND-EXOR expressions.

Each n-variable function has an infinite number of ESOP representations [113]. As a result, many heuristic minimization approaches have been proposed (e.g., Exmin [116] and EXORCISM [93]) to find the ESOP expression with a small number of product terms. In addition, exact ESOP minimization algorithms have been developed (e.g., [60, 114, 115, 137]), which are able to find the ESOP expression with the smallest number of product terms. However, they require a large computation time and are therefore only applicable to small functions.

Several restricted subclasses have been considered in the past, e.g., *positive polarity Reed-Muller expressions* (PPRM [113]), in which all literals are positive, or *fixed polarity Reed-Muller expressions* (FPRMs, [113]), in which each variable must have the same polarity in all product terms. There are further subclasses (e.g., *kronecker expressions* (KROs) [34] and *pseudo kronecker expressions* [71] (PSDKROs)) and most of them can be defined based on applying the following decomposition rules. An arbitrary Boolean function $f(x_1, x_2, \ldots, x_n)$ can be expanded as

$$f = \bar{x}_i f_{\bar{x}_i} \oplus x_i f_{x_i} \qquad \qquad \text{(Shannon)}$$

$$f = f_{\bar{x}_i} \oplus x_i (f_{\bar{x}_i} \oplus f_{x_i}) \qquad \qquad \text{(positive Davio)}$$

$$f = f_{x_i} \oplus \bar{x}_i (f_{\bar{x}_i} \oplus f_{x_i}) \qquad \qquad \text{(negative Davio)}$$

with *co-factors* $f_{\bar{x}_i} = f(x_1, \ldots, x_{i-1}, 0, \ldots, x_n)$ and $f_{x_i} = f(x_1, \ldots, x_{i-1}, 1, \ldots, x_n)$.

The inclusion relationships between the subclasses of AND-EXOR expressions can be stated as the following: $PPRM \subseteq FPRM \subseteq KRO \subseteq PSDKRO \subseteq ESOP$.

Example 2.6. Consider the truth table of the 1-bit full adder specified in Fig. 2.1b, the output carry function is defined as follows.

$$C_{\text{out}}(x_0, x_1, C_{\text{in}}) = x_0 C_{\text{in}} \vee x_1 C_{\text{in}} \vee x_0 x_1.$$

The PPRM representation is obtained by applying only the positive Davio decomposition for each variable.

$$C_{\text{out}}(x_0, x_1, C_{\text{in}}) = x_0 C_{\text{in}} \oplus x_1 C_{\text{in}} \oplus x_0 x_1 C_{\text{in}} \oplus x_0 x_1.$$

The FPRM representation is obtained by applying either the positive or the negative Davio decomposition for each variable. In the following, we apply the positive Davio decomposition to the variables x_0 and C_{in} while for the variable x_1, we apply the negative Davio decomposition.

$$C_{\text{out}}(x_0, x_1, C_{\text{in}}) = C_{\text{in}} \oplus x_0 \oplus x_0 \bar{x}_1 \oplus C_{\text{in}} \wedge \bar{x}_1 \oplus C_{\text{in}} \wedge x_0 \bar{x}_1.$$

The optimal ESOP expression with minimal product terms is either

$$C_{\text{out}}(x_0, x_1, C_{\text{in}}) = C_{\text{in}} x_0 \oplus \bar{C}_{\text{in}} x_1 \oplus \bar{x}_0 x_1$$

or

$$C_{\text{out}}(x_0, x_1, C_{\text{in}}) = C_{\text{in}} x_0 \oplus C_{\text{in}} x_1 \oplus x_0 x_1.$$

2.4   Boolean Satisfiability and SAT Modulo Theory

The *conjunctive normal form* (CNF) consists of a conjunction of clauses. A clause is a disjunction of literals and each literal is a propositional variable or its negation.

The *Boolean satisfiability* (SAT) problem is defined as follows.

Definition 2.2 (Boolean Satisfiability). Let h be a Boolean function in CNF, i.e., a product-of-sum representation. Then, the SAT problem is to find an assignment for the variables of h such that h evaluates to 1 or to prove that such an assignment does not exist.

Example 2.7. Let $F = (x_1 \vee x_2) \wedge (x_1 \vee \bar{x}_3) \wedge (\bar{x}_2 \vee x_3) \wedge (\bar{x}_2 \vee \bar{x}_4)$, then $x_1 = 1$, $x_2 = 0$, $x_3 = 0$, and $x_4 = 0$ is a satisfying assignment for F. The value of x_1 ensures that the first two clauses become satisfied while the value of x_2 ensures that the remaining clauses become satisfied.

SAT is one of the central $\mathcal{NP}$-complete problems. It was the first known $\mathcal{NP}$-complete problem that was proven by Cook in 1971 [30]. Despite this proven complexity today there are SAT algorithms which solve many practical problem instances very fast. Application domains are for example automatic test pattern generation [70, 123], logic synthesis [156], diagnosis [126], and verification [19, 29, 102]. For SAT solving several (backtracking) algorithms have been proposed [35, 36, 45, 72, 95].

The basic search procedure to find a satisfying assignment is shown in Fig. 2.6 and follows the structure of the *Davis Putnam Logemann Loveland* (DPLL) algorithm [35, 36]. The description follows the implementation of the procedure in modern SAT solvers. While there are free variables left (c), a decision is made (e) to assign a value to one of these variables. Then, implications are determined due to the last assignment by BCP (f). This may cause a conflict (g) that is analyzed. If the conflict can be resolved by undoing assignments from previous decisions, backtracking is done (h). Otherwise, the instance is unsatisfiable (i). If no further decision can be done, i.e., a value is assigned to all variables and this assignment did not cause a conflict, the CNF is satisfied (d).

Due to the tremendous improvements in the recent past, several researchers investigated the combination of SAT solvers with decision procedures for decidable theories resulting in *SAT modulo theories* (SMT) [21, 44]. An SMT solver integrates a Boolean SAT solver with other solvers for specialized theories. SMT is about checking the satisfiability of first-order formulas containing operations from various theories such as bit-vectors, arithmetic, arrays, uninterpreted functions, and recursive data-types. Here the SAT solver works on an abstract representation (also in CNF) of the problem and steers the overall search process, while each (partial) assignment of this representation has to be validated by the theory solver for the theory constraints. Thus, advanced SAT techniques together with specialized theory solvers can be utilized.

SMT solvers are available to handle complex formulas such as Z3 [37], Math-SAT [24], Boolector [23], and Yices [44].

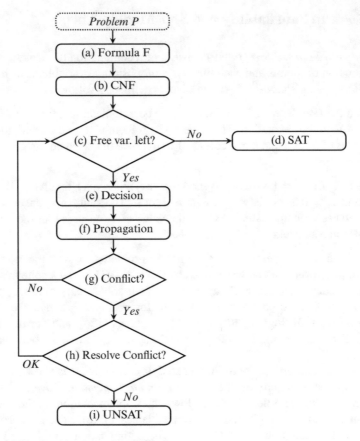

Fig. 2.6 DPLL algorithm in modern SAT solvers

Example 2.8. Let $F = (x_2 + x_3 = -1) \wedge (x_3 > 3x_4) \wedge (x_1 + x_3 < 1) \wedge (x_1 + x_2 > 0)$ be a conjunction of four clauses that contain formulas over theories. Then, $x_1 = 1$, $x_2 = 1$, $x_3 = -2$, and $x_4 = -1$ is a satisfying assignment for F. The values of x_2 and x_3 ensure that the first clause becomes satisfied while x_4 ensures this for the second clause and the assigned value of x_1 asserts the satisfiability of the remaining two clauses.

2.5 Reversible Logic

In contrast to classical design, reversible computation is restricted to operations which can be performed in an invertible fashion. This reversibility builds the basis for many emerging technologies that may replace or at least enhance classical computer chips. Examples include applications in the domain of

Fig. 2.7 Examples of
Boolean functions. (**a**) f_1:
Irreversible function. (**b**) f_2:
Reversible function

a

x_1	x_2	x_3	y_1	y_2	y_3
0	0	0	0	0	0
0	0	1	1	0	0
0	1	0	0	1	0
0	1	1	**0**	**1**	**1**
1	0	0	**0**	**1**	**1**
1	0	1	1	0	1
1	1	0	1	1	1
1	1	1	1	1	0

b

x_1	x_2	x_3	y_1	y_2	y_3
0	0	0	0	0	0
0	0	1	0	1	0
0	1	0	1	0	0
0	1	1	1	0	1
1	0	0	0	0	1
1	0	1	0	1	1
1	1	0	1	1	0
1	1	1	1	1	1

- *Low power computation*, where the fact that no information is lost in reversible computation can be exploited (cf. [17, 18, 69]),
- *Adiabatic circuits*, a special low power technology that reversible circuits are particularly suited for (cf. [100]), and
- *Quantum computation*, which enables the solution of many relevant problems significantly faster than with classical circuits (cf. [98]).

To let the reader be familiar with the concept of reversibility, we introduce in the following the basics on reversible logic. The first part presents the properties of reversible functions. The second part defines the common reversible gates that are used in this book. Finally, reversible circuits and their characteristics are given with examples.

2.5.1 Reversible Function

Definition 2.3 (Reversible Function). A function $f \in \mathscr{B}_{n,n}$ is called *reversible* if f is bijective, i.e., if each input pattern is uniquely mapped to a corresponding output pattern and vice versa. Otherwise, it is called *irreversible*. Clearly, if f is reversible, then its number of inputs is equal to the number of outputs.

In other words, each reversible function $f \in \mathscr{B}_{n,n}$ is a bijection that performs a permutation of the set of input patterns. Therefore, a reversible function $f \in \mathscr{B}_{n,n}$ can be represented by a *permutation* of $\{0, 1, \ldots, 2^n - 1\}$, i.e.

$$\pi_f \stackrel{\text{def}}{=} (\text{nat}(f(0, \ldots, 0, 0)), \ldots, \text{nat}(f(1, \ldots, 1, 1))) \tag{2.2}$$

where $\text{nat} : \mathbb{B}^n \to \{0, 1, \ldots, 2^n - 1\}$ maps a bit-vector to its natural number representation. Further, f can be represented as a $2^n \times 2^n$ Boolean *permutation matrix* Π_f where

$$\left(\Pi_f\right)_{r,c} \stackrel{\text{def}}{=} \left(\pi_f(c) \equiv r\right). \tag{2.3}$$

Example 2.9. The function f_1 presented in Fig. 2.7a is irreversible. Despite that the number of inputs is equal to the number of outputs, there is no unique input-output mapping (e.g., both inputs 011 and 100 map to the same output 011). In contrast, the function f_2 outlined in Fig. 2.7b is reversible, since each input pattern maps to a unique output pattern and the number of inputs is equal to the number of outputs.

The output permutation after applying f_2 is computed in the following:

$$\pi_{f_2} = (\text{nat}(f(000)), \text{nat}(f(001)), \ldots, \text{nat}(f(110), \text{nat}(f(111))))$$

$$= (\text{nat}(000), \text{nat}(010), \ldots, \text{nat}(110), \text{nat}(111))$$

$$= (0, 2, 4, 5, 1, 3, 6, 7)$$

The permutation matrix is in the following:

		Inputs							
		000	001	010	011	100	101	110	111
000		1	0	0	0	0	0	0	0
001		0	0	0	0	1	0	0	0
010		0	1	0	0	0	0	0	0
011	Outputs	0	0	0	0	0	1	0	0
100		0	0	1	0	0	0	0	0
101		0	0	0	1	0	0	0	0
110		0	0	0	0	0	0	1	0
111		0	0	0	0	0	0	0	1

$\Pi_{f_2} =$

An irreversible function can be embedded into a reversible specification by adding extra variables to achieve a bijective function. An embedding is not unique and the choice of embedding can have a very significant effect on the number of the variables of the resulting function [89, 136].

Assume $f = (f_1, \ldots, f_m) \in \mathcal{B}_{n,m}$ and $g = (g_1, \ldots, g_{m'}) \in \mathcal{B}_{n,m'}$, where $m' \geq m$. We write $f = g|_m$ in case $f_i(X) = g_i(X)$ for each $X \in \mathbb{B}^n$ and each $i \in \{1, \ldots, m\}$ and say that f is the *m-projection* of g.

Definition 2.4 (Embedding [136]). A function $g \in \mathcal{B}_{n,m+k}$ *embeds* $f \in \mathcal{B}_{n,m}$, if g is injective and $f \equiv g|_m$. The function g is called an *embedding* and the additional k outputs of g are referred to as **garbage outputs**. Furthermore, c additional input variables might be added such that $n + c = m + k$ in order to obtain a reversible function for the embedding g. The additional c inputs are referred to as **constant inputs**.

More precisely, given an m-output irreversible function f on n variables, a *reversible* function g with $m + k$ outputs is determined such that g agrees with f on the first m components. Then, bijectivity can readily be achieved, e.g., by adding additional inputs such that f evaluates to its original values in case these inputs are assigned the constant value 0 and each output pattern that is not in the image of g is arbitrarily distributed among the new input patterns.

Let $\mu(f)$ denote the number of occurrences of the most frequent output pattern in the truth table of a. Then, $\ell(f) \overset{\text{def}}{=} \lceil \log_2 \mu(f) \rceil$ is the minimum number of garbage outputs (denoted by k) required to convert an irreversible function to a reversible function. Thus, if $k = \ell(f)$, then the embedding g is called *optimal*. Different algorithms that perform an embedding of irreversible functions based on their truth table description have been proposed in the past, e.g., [76, 89, 136].

Example 2.10. The 1-bit full adder specified in Fig. 2.8a is a Boolean function $f \in \mathcal{B}_{3,2}$. It can be embedded into a function $g \in \mathcal{B}_{3,4}$ which is illustrated in Fig. 2.8b. The most frequent output pattern in f is 01. This pattern is repeated three times in rows 2, 3, and 5. Then, $\mu(f) = 3$. As can be seen, the number of additional garbage outputs are $k = 2$ and $\ell(f) = \lceil \log_2 \mu(f) \rceil = 2 = k$, hence the embedding g is optimal. Two additional constant inputs are added to ensure the reversibility of the final embedded functional, i.e., the same number of inputs and outputs. The result of adding a constant input for the embedding function g shown in Fig. 2.8b is given in Fig. 2.8c.

a

x_1 x_2 C_{in}	C_{out} S
0 0 0	0 0
0 0 1	0 1
0 1 0	0 1
0 1 1	1 0
1 0 0	0 1
1 0 1	1 0
1 1 0	1 0
1 1 1	1 1

b

x_1 x_2 C_{in}	C_{out} S g_1 g_2
0 0 0	0 0 0 0
0 0 1	0 1 0 0
0 1 0	0 1 0 1
0 1 1	1 0 0 1
1 0 0	0 1 1 1
1 0 1	1 0 1 1
1 1 0	1 0 1 0
1 1 1	1 1 1 0

c

C x_1 x_2 C_{in}	C_{out} S g_1 g_2
0 0 0 0	0 0 0 0
0 0 0 1	0 1 0 0
0 0 1 0	0 1 0 1
0 0 1 1	1 0 0 1
0 1 0 0	0 1 1 1
0 1 0 1	1 0 1 1
0 1 1 0	1 0 1 0
0 1 1 1	1 1 1 0
1 0 0 0	1 0 0 0
1 0 0 1	1 1 0 0
1 0 1 0	1 1 0 1
1 0 1 1	0 0 0 1
1 1 0 0	1 1 1 1
1 1 0 1	0 0 1 1
1 1 1 0	0 0 1 0
1 1 1 1	0 1 1 0

Fig. 2.8 Embedding of the 1-bit full adder function. (**a**) 1-bit full adder function. (**b**) Garbage outputs. (**c**) Constant input

2.5.2 Reversible Gates

Reversible functions in $\mathscr{B}_{n,n}$ can be realized by reversible circuits that consist of at least n lines and are constructed as circuits of reversible gates that belong to a certain gate library. Prominent reversible logic gates are Toffoli [139], Fredkin [49], and Peres [101]. In the case where a reversible gate library contains a smallest complete set of reversible gates which can be used to implement any reversible function, the library is called a *universal* gate library. The most common universal gate library consists of *Toffoli gates* [139], and *single-target gates* [40, 140]. In the following, we define each of these libraries.

Definition 2.5 (Single-Target Gate). Given a set of variables $X = \{x_1, \ldots, x_n\}$, a *single-target* (ST) gate $T_g(C, t)$ with *control lines* $C = \{x_{i_1}, \ldots, x_{i_k}\} \subset X$, a *target line* $t \in X \setminus C$, and a *control function* $g \in \mathscr{B}_k$ inverts the variable on the target line, if and only if $g(x_{i_1}, \ldots, x_{i_k})$ evaluates to true. All other variables remain unchanged. If the definition of g is obvious from the context, it can be omitted from the notation $T_g(C, t)$.

Example 2.11. Figure 2.9 shows the diagrammatic representation of a single-target gate based on Feynman's notation. The target line is denoted with the symbol $\oplus$.

Given n Boolean variables, the maximum number of variables that can be involved in a controlling function of an ST gate is $n - 1$ and the remaining variable would be used to compute the target for that gate. A controlling function g of an ST gate denotes an arbitrary Boolean function of $n - 1$ Boolean variables. It is well known that the number of possible Boolean functions with $n - 1$ variables is $2^{2^{n-1}}$, hence with n different variables taken each as a target line we get $n \cdot 2^{2^{n-1}}$ possible ST gates as depicted in Table 2.1.

Definition 2.6 (Toffoli Gate). *Mixed-polarity multiple-control Toffoli* (MPMCT) gates are a subset of the single-target gates in which the control function g can be represented with one product term $g : C = (x_{i_1}, x_{i_2}, \ldots, x_{i_k}) \mapsto \bigwedge_{k=i}^{j} x_i^p$. *Multiple-control Toffoli* gates (MCT) are a subset from MPMCT gates in which the product terms can only consist of positive literals.

For drawing circuits, we follow the established conventions of using the symbol solid black circles (•) to indicate positive controls and white circles (○) to indicate

Fig. 2.9 Single-target gate representation

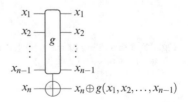

Table 2.1 Number of reversible gates using n lines

Gate type	Formula	Number of gates
NOT	$n \cdot \binom{n-1}{0}$	n
CNOT	$n \cdot \binom{n-1}{1}$	$n \cdot (n-1)$
Toffoli	$n \cdot \binom{n-1}{2}$	$\frac{n \cdot (n-1) \cdot (n-2)}{2}$
NCT	$n \cdot \sum_{i=0}^{2} \binom{n-1}{i}$	$n \cdot \left(\frac{n \cdot (n-1)}{2} + 1 \right)$
MCT	$n \cdot \sum_{i=0}^{n-1} \binom{n-1}{i}$	$n \cdot 2^{n-1}$
MPMCT	$n \cdot \sum_{i=1}^{n-1} \binom{n-1}{i} \cdot 2^{i}$	$n \cdot 3^{n-1}$
ST	$n \cdot |\mathscr{B}_{n-1}|$	$n \cdot 2^{2^{n-1}}$
Arbitrary	$P_{2^n}^{2^n}$	$(2^n)!$

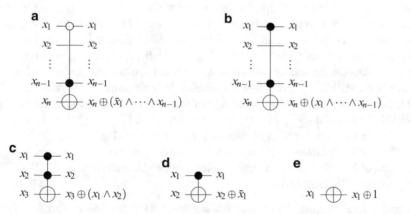

Fig. 2.10 Toffoli library. (**a**) MPMCT gate. (**b**) MCT gate. (**c**) Toffoli gate. (**d**) CNOT gate. (**e**) NOT gate

negative controls. Also, MPMCT gates can be represented as a diagrammatic representation as a single-target gate representation.

Example 2.12. Figure 2.10a shows an MPMCT gate with mixed polarity control lines over the set of variables $X = \{x_1, \ldots, x_n\}$, while Fig. 2.10b shows an MCT gate with positive control lines. The reaming figures present the Toffoli gate, the CNOT gate, and the NOT gate.

An MCT gate with two controls is called *Toffoli* gate, while when it has only one control it is called a *CNOT* gate, and in case it has no control it is named a *NOT* gate. When a library contains only gates form the gate set type { NOT, CNOT, Toffoli }, then this library is referred as *NCT library*. It is called the *MCT library* when only MCT gates are used, otherwise it is called the *MPMCT library*.

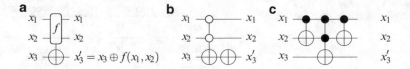

Fig. 2.11 ST gate with a controlling function ($\vee$) given by Fig. 2.1b. (**a**) ST gate. (**b**) MPMCT circuit. (**c**) MCT circuit gate

With n different variables, the total number of the conceivable NOT gates is equal to n as depicted in Table 2.1 since each variable can have a NOT gate. Furthermore, the number of all the possible CNOT and Toffoli gates are $n \cdot (n - 1)$ and $n \cdot \binom{n-1}{2}$, respectively. This is calculated as the following: in a set X, the total number of subset having k elements is the sum of all the combinations with k elements from that set. If the set has $n - 1$ elements, the number of k-combinations is equal to the binomial coefficient $\binom{n-1}{k}$. Assuming that we have as a starting point n different sets, hence we have n multiplied by the total number of combinations. Therefore the total number of gates that an NCT library and MCT library is the total number of binomial coefficients with $k = 0, 1, 2$ and $k = 0, 1, 2, \ldots, n - 1$, respectively where k denotes the number of controls. Consider the possible number of MPMCT gates in an MPMCT library over n variables. It is the same as the number of MCT gates except that we add the possibility of negative controls. Hence, the number of MPMCT gates is $n \cdot \sum_{i=1}^{n-1} \binom{n-1}{i} \cdot 2^i$ and given $\sum_{i=1}^{n} \binom{n}{i} \cdot x^i \cdot y^{n-i} = (x + y)^n$, one can deduct that this the total number of possible MPMCT gates is $n \cdot 3^{n-1}$.

Since its controlling function is Boolean, an ST gate can be expressed in terms of a circuit of MPMCT or MCT gates, which can be obtained from an ESOP or PPRM expression [113], respectively.

Example 2.13. Figure 2.11a shows a single-target gate. The controlling function of this gate is the ($\vee$) function specified in Fig. 2.1b. The equivalent MPMCT gate circuit is given in Fig. 2.11b, while Fig. 2.11c shows its correspondent MCT gate circuit.

2.5.3   Reversible Circuits

A reversible circuit is defined as follows.

Definition 2.7 (Reversible Circuit [107]). A combinational reversible circuit is an acyclic combinational logic circuit in which all gates are reversible, and are interconnected without explicit fan-outs and loops.

Boolean functions can be synthesized to a reversible circuit after embedding them to reversible functions. Therefore, in general a reversible circuit contains n inputs with p primary inputs and c constant inputs with $p + c = n$. At the output side, there are m primary outputs and k garbage outputs with $k + m = n$. Figure 2.12

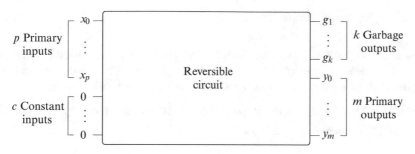

Fig. 2.12 Reversible circuit structure inspired from [107]

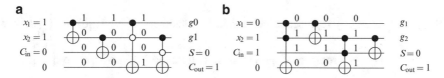

Fig. 2.13 Reversible circuits for 1-bit full adder as specified in Fig. 2.8. (**a**) MPMCT library based circuit. (**b**) MCT library based circuit

depicts the general structure of a reversible circuit. Note that when the function is bijective, there are neither constant inputs nor garbage outputs. This follows from [122], where it has been shown that any reversible function $f \in \mathscr{B}_{n,n}$ can be realized by a reversible circuit with n lines when using MCT gates. This means that it is not necessary to add any temporary lines (referred to as *ancilla*) to realize the circuit.

Definition 2.8 (Ancilla). Let G be a reversible or quantum circuit. An *ancilla* is an additional line whose input is arbitrary or set to a constant value and is used in a way throughout the circuit to compute intermediate results.

Reversible functions can be realized by reversible circuits that consist of at least n lines and are constructed as circuits of reversible gates that belong to a certain gate library (e.g., NCT, MCT, or MPMCT library [139]).

Example 2.14. Figure 2.13a shows different MPMCT gates in a circuit forming a reversible circuit that realizes the embedded 1-bit full adder specified in Fig. 2.8b. The annotated values demonstrate the computation of the gates for a given input assignment, while Fig. 2.13b depicts the equivalent circuit based on MCT gates.

2.6 Quantum Computation

In this section, we present briefly the fundamental concepts of quantum states and systems. Then, we introduce quantum libraries and several examples of quantum physical realizations.

2.6.1 Quantum Systems

Quantum computation [98] provides a new paradigm of computation based on the so-called *qubits* instead of bits. In contrast to Boolean logic, *qubits* do not only represent Boolean 0's and Boolean 1's, but also the superposition of both.

Definition 2.9 (Qubit). A *qubit* $|\varphi\rangle$ (bra-ket notation introduced by Paul Dirac) is a vector $\begin{bmatrix} a_0 \\ a_1 \end{bmatrix}$ where $a_0,\ a_1 \in \mathbb{C}$ such that

$$|a_0|^2 + |a_1|^2 = 1. \tag{2.4}$$

If $a_0 = 1$, then $|\varphi\rangle$ represents the classical 0, denoted $|0\rangle$ with $|0\rangle \equiv \begin{bmatrix} 1 \\ 0 \end{bmatrix}$, and if $a_1 = 1$, then $|\varphi\rangle$ represents the classical 1, denoted $|1\rangle$, with $|1\rangle \equiv \begin{bmatrix} 0 \\ 1 \end{bmatrix}$. Any state of a *qubit* may also be written as

$$|\varphi\rangle = a_0|0\rangle + a_1|1\rangle. \tag{2.5}$$

Using quantum computation and *qubits* in superposition, functions can be evaluated with different possible input assignments in parallel. It is not possible to obtain the current state of a *qubit*. Instead, if a *qubit* is measured, either 0 or 1 is returned depending on a respective probability, i.e., 0 is returned with a probability of $|a_0|^2$ and 1 is returned with a probability of $|a_1|^2$. After the measurement, the state of the qubit is destroyed.

The state of a quantum system with $n \geq 1$ qubits is given by an element of the tensor product of the single qubit spaces, i.e., a linear combination of the 2^n tensor states $|0\cdots0\rangle$, $|0\cdots1\rangle$, ..., and $|1\cdots1\rangle$, which are the tensor products of basis states. Consequently, a quantum state is represented as a normalized vector of length 2^n (called the state vector), whose components denote the amplitude for each tensor state. The state of an n-qubit quantum system can be represented as the following equation:

$$|\psi\rangle = \sum_{X\in\{0,1\}^n} a_X|X\rangle \tag{2.6}$$

with

$$\sum_{X\in\{0,1\}^n} |a_X|^2 = 1 \tag{2.7}$$

Example 2.15. Consider two states of a 1-qubit quantum system $|\psi\rangle = a|0\rangle + b|1\rangle$ and $|\varphi\rangle = c|0\rangle + d|1\rangle$. The state space of the composite system $|\Phi\rangle$ is the tensor product of the state spaces of the component systems $|\psi\rangle$ and $|\varphi\rangle$. Then:

$$|\Phi\rangle = |\psi\rangle \otimes |\phi\rangle$$
$$= (a|0\rangle + b|1\rangle) \otimes (c|0\rangle + d|1\rangle)$$
$$= ac(|0\rangle \otimes |0\rangle) + ad(|0\rangle \otimes |1\rangle) + bc(|1\rangle \otimes |0\rangle) + bd(|1\rangle \otimes |1\rangle)$$

with $|0\rangle \otimes |0\rangle = \begin{bmatrix} 1 \\ 0 \end{bmatrix} \otimes \begin{bmatrix} 1 \\ 0 \end{bmatrix} = \begin{bmatrix} 1 \begin{bmatrix} 1 \\ 0 \end{bmatrix} \\ 0 \begin{bmatrix} 1 \\ 0 \end{bmatrix} \end{bmatrix} = \begin{bmatrix} 1 \\ 0 \\ 0 \\ 0 \end{bmatrix} = |00\rangle$. Similarly calculated,

the tensor products $|0\rangle \otimes |1\rangle$, $|0\rangle \otimes |1\rangle$, and $|0\rangle \otimes |1\rangle$ are equal to $|01\rangle$, $|10\rangle$, and $|11\rangle$, respectively. Hence,

$$|\Phi\rangle = ac|00\rangle + ad|01\rangle + bc|10\rangle + bd|11\rangle.$$

A state of a n-qubit quantum system is referred to as an entangled state when it cannot be written as any combination of the states of the individual qubits. In other words, the state cannot be represented as the tensor product of n single qubits.

Definition 2.10 (Entanglement). Let $|\psi\rangle$ denote a composite state. Then $|\psi\rangle$ is said to be entangled if $|\psi\rangle$ cannot be expressed in the form:

$$|\psi\rangle = |\psi_1\rangle \otimes |\psi_2\rangle \tag{2.8}$$

otherwise is said to be separable.

Example 2.16. The bell state [16] $|\beta\rangle = \frac{1}{\sqrt{2}}|00\rangle + |11\rangle$ is an entangled state. Also any state $|\phi\rangle = a|00\rangle + b|11\rangle$ is an entangled state provided that $a \neq 0$ and $b \neq 0$. While the state $|\psi\rangle = a|10\rangle + b|11\rangle$ is a separable state since $|\psi\rangle = a|10\rangle + b|11\rangle = |1\rangle \otimes (a|0\rangle + b|1\rangle)$.

Definition 2.11 (Unitary Matrix). A matrix U is *unitary* if $U^\dagger U = UU^\dagger = I$ where $U^\dagger = (U^*)^T$ is the conjugate transpose of U.

Example 2.17. Consider the matrix $U = \begin{bmatrix} 1 & 0 \\ 0 & i \end{bmatrix}$.

Its conjugate transpose is $U^\dagger = \begin{bmatrix} 1 & 0 \\ 0 & -i \end{bmatrix}$. Since $UU^\dagger = \begin{bmatrix} 1 & 0 \\ 0 & 1 \end{bmatrix}$ and $U^\dagger U = \begin{bmatrix} 1 & 0 \\ 0 & 1 \end{bmatrix}$, hence U is a unitary matrix.

Following the theory of quantum mechanics, the evolution of a quantum system due to a quantum operation can be described by a unitary transformation matrix U. Here, the columns correspond to the output state vectors that result when applying the respective operation to the tensor states as inputs. Thus, the entry u_{ij} of the matrix describes the mapping from the input tensor state $|j_i\rangle$ to the output tensor state $|i_j\rangle$.

Example 2.18. Consider the bell state [16] which is a 2-qubit quantum system defined by $|\beta\rangle = \frac{1}{\sqrt{2}}|00\rangle + |11\rangle$. In the following, we determine the unitary transformation matrix U that represents the 2-qubit quantum system $|\beta\rangle$:

$$|\beta\rangle = \frac{1}{\sqrt{2}}|00\rangle + |11\rangle$$

$$= \frac{1}{\sqrt{2}}|00\rangle + 0|01\rangle + 0|10\rangle + 1|11\rangle$$

$$= \frac{1}{\sqrt{2}}\begin{bmatrix}1\\0\\0\\0\end{bmatrix} + 0\begin{bmatrix}0\\1\\0\\0\end{bmatrix} + 0\begin{bmatrix}0\\0\\1\\0\end{bmatrix} + 1\begin{bmatrix}0\\0\\0\\1\end{bmatrix}, \text{ hence,}$$

$$U = \begin{bmatrix} \frac{1}{\sqrt{2}} & 0 & 0 & 0 \\ 0 & 0 & 0 & 0 \\ 0 & 0 & 0 & 0 \\ 0 & 0 & 0 & 1 \end{bmatrix}$$

2.6.2 Quantum Libraries

Since most of the known quantum algorithms include a large Boolean component (e.g., the database in Grover's search algorithm and the modulo exponentiation in Shor's algorithm), the design of these components is often conducted by a two-stage approach: (1) realizing the desired functionality as a reversible circuit and (2) mapping the resulting circuit to a functionally equivalent quantum circuit.

The *technology mapping* is the process where a reversible circuit is converted to a quantum circuit using a selected set of gates from a technology library. The technology mapping is also called decomposition or transformation. The currently known mappings of MCT/MPMCT gates into quantum circuits using ancillae have been introduced in [15, 91, 98]. In the scope of this book, we are considering circuits that realize pure Boolean functionality but still need to be realized using quantum gates in order to embed them into quantum algorithms such as Deutsch-Josza [41], Grover [56], or Shor [125, 141].

Definition 2.12 (Quantum Gate). In general, a quantum gate acting on n qubits represents a $2^n \times 2^n$ unitary matrix [98]. A *single qubit gate* $U(t)$ over the inputs $X = \{x_1, \ldots, x_n\}$ consists of a single-target line $t \in X$, while a *two qubit gate* (also denoted as *single-control gate*) $U(c, t)$ comprises, in addition, a single control line $c \in X$ with $t \neq c$. Quantum gates are also called *elementary* gates.

Different libraries of quantum gates have thereby been introduced for this purpose. In the literature, the *NCV library* [15] is frequently applied. This library is defined in the following:

Definition 2.13 (NCV Library). We consider the gate library $\{NOT, CNOT, V, V^\dagger\}$ as a universal quantum gate set for implementing any reversible function. The V and $V^\dagger$ gate are both the square root of the $CNOT$ gate since two adjacent identical V, or $V^\dagger$, gates are equivalent to a CNOT gate. The $V^\dagger$ gate is the inverse of the V gate (Table 2.2).

Table 2.2 Gate definitions

Type	Symbol	Matrix	Diagram
NOT	N	$\begin{bmatrix} 0 & 1 \\ 1 & 0 \end{bmatrix}$	
CNOT	C	$\begin{bmatrix} 1&0&0&0 \\ 0&1&0&0 \\ 0&0&0&1 \\ 0&0&1&0 \end{bmatrix}$	
Controlled V	V	$\begin{bmatrix} 1&0&0&0 \\ 0&1&0&0 \\ 0&0&\frac{1+i}{2}&\frac{1-i}{2} \\ 0&0&\frac{1-i}{2}&\frac{1+i}{2} \end{bmatrix}$	
Controlled V^{-1}	$V^\dagger$	$\begin{bmatrix} 1&0&0&0 \\ 0&1&0&0 \\ 0&0&\frac{1-i}{2}&\frac{1+i}{2} \\ 0&0&\frac{1+i}{2}&\frac{1-i}{2} \end{bmatrix}$	
Hadamard	H	$\frac{1}{\sqrt{2}}\begin{bmatrix} 1 & 1 \\ 1 & -1 \end{bmatrix}$	
Z-gate	Z	$\begin{bmatrix} 1 & 0 \\ 0 & -1 \end{bmatrix}$	
Phase	S	$\begin{bmatrix} 1 & 0 \\ 0 & i \end{bmatrix}$	
Phase^{-1}	$S^\dagger$	$\begin{bmatrix} 1 & 0 \\ 0 & -i \end{bmatrix}$	
T-gate	T	$\begin{bmatrix} 1 & 0 \\ 0 & e^{\frac{i\pi}{4}} \end{bmatrix}$	
T-gate^{-1}	$T^\dagger$	$\begin{bmatrix} 1 & 0 \\ 0 & e^{\frac{-i\pi}{4}} \end{bmatrix}$	

Table 2.3 Possible qubit
values for NCV gates

x	$NOT(x)$	$V(x)$	$V^+(x)$
0	1	v_0	v_1
v_0	v_1	1	0
1	0	v_1	v_0
v_1	v_0	0	1

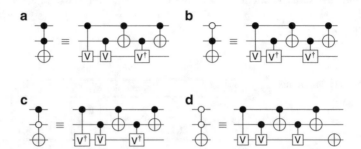

Fig. 2.14 NCV quantum mapping of different Toffoli gates. (**a**) Toffoli with two positive controls.
(**b**) Toffoli with a negative and a positive control. (**c**) Toffoli with a positive and a negative control.
(**d**) Toffoli with a negative controls

If circuits with Boolean inputs use NCV gates only, the value of each qubit at each
stage of the circuit is restricted to one of $\{0, v_0, 1, v_1\}$, where $v_0 = \frac{1+i}{2}\binom{1}{-i}$ and
$v_1 = \frac{1+i}{2}\binom{-i}{1}$. The NOT, V, and V^+ operations over these four-valued logic are
given in Table 2.3. As shown, NOT is a complement operation, V is a cycle, and V^+
is the inverse cycle.

Example 2.19. Figure 2.14 shows quantum circuits consisting of NCV gates. Each
of these circuits depicts one of the minimal realizations of a Toffoli gate with respect
to the polarity assignments for its controls.

Recently, there has been particular interest in quantum circuits composed of
Clifford$+T$ gates. This is motivated by the importance of fault tolerance in quantum
computations [144]. Unlike the NCV library, the Clifford$+T$ library is universal for
quantum computation, i.e., every quantum system can be realized by it [22] as well
as its gates can be implemented in a fault-tolerant way [26, 61].

Definition 2.14 (Clifford$+T$ **Library**). We consider the library $\{H, Z, S, T, CNOT\}$
as the universal gate set for implementing any quantum system. Note that the $S^\dagger$,
$T^\dagger$, and NOT gates can be implemented with ZS, ZST, and HZH, respectively. The
S and $S^\dagger$ gates are square roots of the Z gate. Similarly, the T and $T^\dagger$ gates are given
by matrices that are the fourth root of the Z gate (See Table 2.2).

A *single qubit gate* $G(t)$ over the inputs $X = \{x_1, \ldots, x_n\}$ consists of a single-
target line $t \in X$, while a CNOT gate $G(c, t)$ comprises, in addition, a single control
line $c \in X$ with $t \neq c$.

Fig. 2.15 Clifford + T
quantum mapping of different
Toffoli gates. (**a**) Toffoli with
two positive controls. (**b**)
Toffoli with a negative and a
positive control. (**c**) Toffoli
with a positive and a negative
control. (**d**) Toffoli with a
negative controls

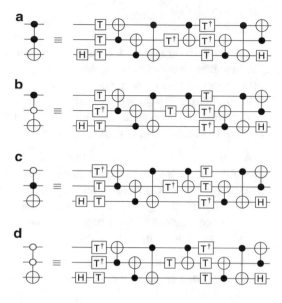

Table 2.4 Number of quantum gates using n lines

Gate/library type	Formula	Number of gates
Noncontrolled gate[a]	$n \cdot \binom{n-1}{0}$	n
Single-control gate[a]	$n \cdot \binom{n-1}{1}$	$n \cdot (n-1)$
NCV	$n \cdot \binom{0}{n-1} + 3n \cdot \binom{n-1}{1}$	$n \cdot (3n-2)$
Clifford + T	$6n \cdot \binom{0}{n-1} + n \cdot \binom{n-1}{1}$	$n \cdot (n+6)$
Arbitrary	∞	∞

[a]NCV or Clifford + T gate

Example 2.20. Figure 2.15 shows the quantum realizations consisting of
Clifford + T gates. These circuits represent the optimal realizations of Toffoli
gates with all possible polarities that its controls might take.

The major cost-aware objective in Clifford + T circuits is to minimize the number
of T gates and particularly the T-depth of the circuit due to the fact that the cost of
the fault tolerant implementation of a T gate can exceed the cost of implementing a
Clifford gate by a factor of 100 or more [10].

Table 2.4 outlines the total number of elementary gates that can be implemented
over n variables. For example the NCV library has $n \cdot (3n - 2)$ different quantum
gates while the Clifford + T library contains $n \cdot (n + 6)$ gates, where n is the number
of possible qubits. The number of quantum arbitrary gates is infinite if we consider
the infinite roots of the Z, *NOT*, or *CNOT* gates.

2.6.3   Quantum Circuits

Definition 2.15 (Quantum Circuit). A quantum circuit is formulated as circuits of quantum gates over one or two qubits from a particular library. Unary quantum gates apply its operations at the respective qubit, while two-qubit gates have a *control qubit* in addition. The respective operation at the target qubit is performed if the control qubit is 1. A quantum circuit can be represented by a unitary matrix that is computed as the product of the matrices representing the individual gates.

A quantum circuit can be obtained through two different strategies: (1) a reversible circuit is synthesized, then each MPMCT gate is mapped to its quantum circuits [15, 91], or (2) direct synthesis finds the suitable circuit of quantum gates that can represent a given unitary matrix based on decomposition or exhaustive search. Many algorithms have been presented for NCV library [55, 82] as well as for the Clifford $+ T$ library [51, 65].

Example 2.21. The circuits depicted in Fig. 2.16a, b are the NCV and Clifford $+ T$ quantum circuits, respectively, for the reversible circuit shown in Fig. 2.13b realizing a 1-bit full adder.

When a resulting mapped circuit does not produce any entangled state, it is called a *Semi-Classical Quantum Circuit*.

Definition 2.16 (Semi-Classical Quantum Circuit). A Semi-Classical Quantum Circuit (SCQC) is a quantum circuit in which if all the initial input quantum states of the circuit are in the base states $|0\rangle$ or $|1\rangle$ (classical values), the quantum states of all gate controls in the circuit are also in the base states $|0\rangle$ or $|1\rangle$ [154]. Entanglement does not occur in SCQCs as long as their inputs are initialized to classical values.

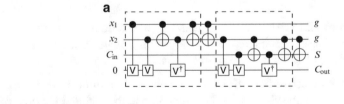

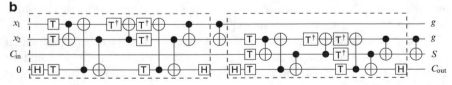

Fig. 2.16 Mapping of the 1-bit full adder reversible circuit in Fig. 2.13b. (**a**) NCV based quantum circuit. (**b**) Clifford $+ T$ based quantum circuit

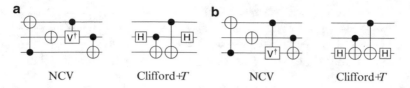

Fig. 2.17 Examples of entangled and SCQC quantum circuits. (**a**) Entangled circuits. (**b**) SCQC circuits

A quantum circuit is called an entangled circuit when it generates an entangled state for some given binary input state. If the state of a qubit is either v_0 or v_1, and is applied to the control of a two-qubit gate, an entangled state may occur. Consequently, the resulting output state cannot be represented by the combination of individual qubit states.

Example 2.22. The circuits shown in Fig. 2.17a are entangled circuits. Consider the circuit in Fig. 2.17a. (1), it generates an entangled state for the input state vector $\langle 0, 0, 1 \rangle$ and the resulting output state vector cannot be separable into 3 single-qubit state vectors. The circuits sketched in Fig. 2.17b present two SCQC quantum circuits.

2.7   Cost Metrics for Reversible and Quantum Circuits

To compare the efficiency of different synthesis approaches it is important to evaluate the resulting circuits. Depending on the target application, different metrics are applied to measure the complexity of a given circuit.

2.7.1   Quantum Cost

The *quantum cost* denotes the required effort to transform a reversible circuit to a quantum circuit. The quantum cost depends on the used quantum library, and it is denoted as NCV quantum cost (*NCV-cost*) when the NCV library is used and *T-depth* when the Clifford + *T* library is used.

Definition 2.17 (NCV-Cost). The *NCV-cost* of a quantum circuit is the total number of quantum gates that form the circuit, while it denotes the number of required elementary gates to decompose that circuit to a quantum circuit.

Example 2.23. Consider the NCV based quantum circuit that implements a 1-bit full adder shown in Fig. 2.18a. This circuit is an optimal realization with respect to the NCV-cost. The quantum circuit has an NCV-cost of 6. An equivalent circuit outlined in Fig. 2.16a has an NCV-cost of 12.

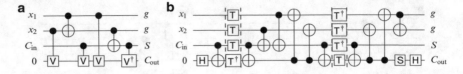

Fig. 2.18 Quantum circuits realizing a 1-bit full adder. (**a**) Optimal NCV circuit. (**b**) Optimal Clifford + T circuit

Table 2.5 Quantum cost for MPMCT gates with c controls

Number of controls (c)	NCT gate-count	NCV-cost ($+2$)[a]	T-depth
0	1	1	0
1	1	1	0
2	1	5	3
3	4	20	12
4	10	40	30
≥ 5	$8(c-3)$[15]	$40(c-3)$	$24(c-3)$

[a]In case all the controls are negative, the NCV-cost is incremented by 2

Definition 2.18 (*T-Depth*). The *T-depth* is the number of T-stages in a quantum circuit where each stage consists of one or more T or $T^\dagger$ gates performed in parallel on separate qubits.

The motivation for that cost measure originates from the fact that the T gate is significantly larger compared to the other gates in the circuit. The *T-count* of a Clifford + T circuit is the total number of T and $T^\dagger$ gates in the circuit, while the *H-count* denotes the total number of H gates in a quantum circuit.

Example 2.24. The Clifford + T optimal circuit with respect to the T-depth of a 1-bit full adder is depicted in Fig. 2.18b. This circuit has a T-depth of 2, *T-count* of 8, and *H-count* of 2, while its equivalent realization sketched in Fig. 2.16b has a T-depth of 6, *T-count* of 14, and *H-count* of 4.

The quantum cost can be directly calculated from a reversible circuit. It is calculated based on the principles proposed originally in [15]. First, each MPMCT gate in a given circuit is mapped to an NCT circuit, then each Toffoli gate in the later circuit is evaluated by its quantum cost. Finally, one can obtain the quantum cost of a reversible circuit by computing the sum of quantum cost for each gate. For more details, the reader is refereed to Sect. 4.1.1.

Table 2.5 reports all the different quantum costs as well as the gate-count of an NCT circuit for each MPMCT gate.

Example 2.25. In Fig. 2.19, the MPMCT based reversible circuit has an NCV-cost of 107 and a T-depth of 63. Based on Table 2.5, we will calculate the quantum cost for each gate, and then the cost for the whole circuit.

Fig. 2.19 MPMCT based
reversible circuit

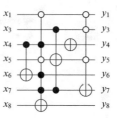

The NCV-cost is obtained in the following:

$$\text{NCV-cost}_{g_1} = 1$$

$$\text{NCV-cost}_{g_2} = 40 \cdot (5 - 3) = 80$$

$$\text{NCV-cost}_{g_3} = 5$$

$$\text{NCV-cost}_{g_4} = 1$$

$$\text{NCV-cost}_{g_5} = 20$$

$$\text{NCV-cost}_G = \sum_{i=1}^{5} \text{NCV} - \text{cost}_{g_i} = 1 + 80 + 5 + 1 + 20 = 107$$

The T-depth is calculated with the same way:

$$T\text{-depth}_{g_1} = 0$$

$$T\text{-depth}_{g_2} = 24 \cdot (5 - 3) = 48$$

$$T\text{-depth}_{g_3} = 3$$

$$T\text{-depth}_{g_4} = 0$$

$$T\text{-depth}_{g_5} = 12$$

$$T\text{-depth}_G = \sum_{i=1}^{5} T\text{-depth}_{g_i} = 0 + 48 + 3 + 0 + 12 = 63$$

2.7.2   Number of Gates

Definition 2.19 (Number of Gates). This metric is related to reversible circuits. It is referred to the total number of gates in a reversible circuit. It is also denoted as *gate-count*.

Example 2.26. The reversible circuit shown in Fig. 2.19 has a number of gates of 5. Figure 2.20a illustrates a reversible circuit with a number of gates of 6.

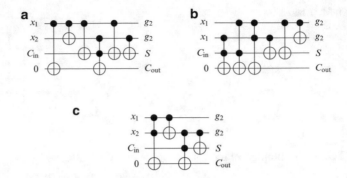

Fig. 2.20 Reversible circuits realizing a 1-bit full adder. (**a**) First realization. (**b**) Second realization. (**c**) Optimal circuit regarding the gate-count

This metric is used to evaluate a given realizations for a reversible function. However, this metric does not give any information about the quantum cost of the corresponding quantum circuit. The minimization based on this metric does not guarantee a minimization of the quantum cost and the inverse is also not true. In fact it is possible that two different realizations for the same reversible function have different gate-count in a strictly increasing order, but the quantum cost for the first realization is more expensive than the second.

Example 2.27. Figure 2.20 shows two different realizations of a 1-bit full adder and a third optimized circuit for both in terms of the number of gates. Although, the first reversible circuit depicted in Fig. 2.20a has more gates (6 gates) than the circuit presented in Fig. 2.20c (4 gates), the first realization is cheaper in terms of quantum cost than the second circuit. More precisely, the NCV-cost and the T-depth of the first circuit are 10 and 3, respectively. The second circuit has an NCV-cost of 12 and a T-depth of 6. Consider the circuits given in Figs. 2.22b and 2.20c. The first circuit has more gates as well as higher quantum cost than the second. Therefore, as mentioned above, minimizing a circuit with respect to the gate-count does not guarantee a minimization on the quantum cost.

2.7.3   Number of Lines

Definition 2.20 (Number of Lines). The number of lines refers to the total number of the input variables or the output variables used in a reversible or a quantum circuit.

Note that if the function to be synthesized is reversible, the number of circuit lines can be equal to the number of the inputs. However, in the case where the Boolean function to be synthesized is irreversible, then additional variables (i.e., constant inputs and garbage outputs) are unavoidable [76]. Therefore, there are several optimization approaches that target the reduction of the number of lines in reversible circuits [148].

Example 2.28. Figure 2.22a shows a reversible circuit with a number of lines of 4, where the number of primary inputs, constants, primary outputs, and garbage outputs are 3, 1, 2, and 2, respectively.

2.7.4   Depth

This metric recognizes whether gates can be applied in parallel which can lead to a reduction in the execution time of a circuit. This cost metric is related to quantum circuits composed by the NCV library or the Clifford $+ T$ library.

Definition 2.21 (Depth). Let $U_i(C_i, t_i)$ and $U_{i+1}(C_{i+1}, t_{j+1})$ be two consecutive quantum gates. These gates can be applied *concurrently* if

$$|C_i \cup C_{i+1} \cup \{t_i, t_{i+1}\}| = |C_i| + |C_{i+1}| + 2.$$

In other words, if the lines used by each gate (both control and target line) are disjoint. Let G be a quantum circuit with k elementary quantum gates, then G can be partitioned into $d \leq k$ subcircuits whose gates can be pairwise applied concurrently. We refer to the minimal d as the *depth* of the circuit.

It is defined as a sub-sequence of gates in a circuit that can be applied in parallel. We assume in this book that two or more gates can operate at the same level if they operate on disjoint qubits and can be grouped together in the circuit.

Example 2.29. The quantum circuit that realizes a 1-bit full adder illustrated in Fig. 2.21 has an NCV-cost of 6. Gates in position 2 and 3 can be applied *concurrently* because they don't share any line together. The same for gates in positions 4 and 5. Hence, the depth of this circuit is 4.

Fig. 2.21 Quantum depth for a 1-bit full adder circuit

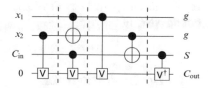

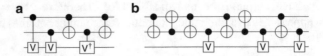

Fig. 2.22 NCV quantum circuits realizing the Toffoli gate given in Fig. 2.10c. (**a**) First realization. (**b**) Second realization

2.7.5 Nearest Neighbor Cost

Many physical realizations of quantum circuits permit limited interaction distance between gate qubits [47]. For example, trapped ions (cf. [58]), liquid nuclear magnetic resonance (cf. [68]), and architectures based on the original Kane model [62] assume that the computations are only performed between adjacent (i.e., nearest neighbor) qubits [47].

Hence, synthesis of quantum circuit with nearest neighbor quantum gates has became an important research branch [27, 59, 63, 96, 107, 151]. This cost is referred to *nearest neighbor cost*.

Definition 2.22 (Nearest Neighbor Cost). Let $U_i(C_i, t_i)$ be quantum gate. The *nearest neighbor cost* (NNC) of U_i is defined $|C_i - t_i - 1|$, i.e., the distance between the control and the target lines of an elementary gate. Let G be a quantum circuit with k elementary quantum gates. Then, the NNC of G is defined as the sum of the NNCs of its gates:

$$\sum_{i=0}^{k} |C_i - t_i - 1| \qquad (2.9)$$

Optimal NNC for a circuit is 0, where all quantum gates are either 1-qubit or 2-qubit gates performed on adjacent qubits. The NNC is also denoted by the interaction cost.

Example 2.30. The quantum circuits given in Fig. 2.22 are possible quantum realizations for a Toffoli gate. The first implementation is an optimal circuit regarding the quantum cost. It has an NCV-cost of 5 but an NNC of 1, while the second circuit represents an optimal circuit with respect to the NNC since it has an NCV-cost of 9 but an NNC of 0.

2.8 Decision Diagrams

2.8.1 Binary Decision Diagrams

A *binary decision diagram* (BDD) is a standard data structure for representing Boolean functions. The general concepts for BDDs are briefly outlined in this section, nevertheless, the reader is referred to the literature for a comprehensive overview [25, 67].

Let $X = x_1, \ldots, x_n$ be the variables of a Boolean function $f \in \mathscr{B}_n$. A BDD representing the function f is a directed acyclic graph with nonterminal vertices N and terminal vertices $T \subseteq \{\boxed{\perp}, \boxed{\top}\}$ where $N \cap T = \emptyset$ and $T \neq \emptyset$. Each nonterminal vertex $v \in N$ is labeled by a variable from X and has exactly two children, low v and high v. The directed edges to these children are called *low-edge* and *high-edge* and are drawn dashed and solid, respectively. A nonterminal vertex v labeled x_i represents a function denoted $\sigma(v)$ given by the *Shannon decomposition* [121]

$$\sigma(v) = \bar{x}_i \sigma(\text{low } v) + x_i \sigma(\text{high } v) \tag{2.10}$$

where $\sigma(\text{low } v)$ and $\sigma(\text{high } v)$ are the functions represented by the children of v with $\sigma(\boxed{\perp}) = 0$

$$\sum_{i=0}^{k} |C_i - t_i - 1|$$

and $\sigma(\boxed{\top}) = 1$.

$$T \subseteq \{\boxed{\perp}, \boxed{\top}\}$$

The BDD has a single start vertex s with $\sigma(s) = f$.

A BDD is *ordered* if the variables of the vertices on every path from the start vertex to a terminal vertex adhere to a specific ordering. Not all of the variables need to appear on a particular path, but a variable can appear at most once on any path. A BDD is *reduced* if there are no two nonterminal vertices representing the same function. Hence, the representation of common sub-functions is shared. In the following, only reduced and ordered BDDs are considered and for briefness referred to as BDDs.

Multiple-output functions can be represented by a single BDD that has more than one start vertex. Common sub-functions that can be shared among the functions decrease the overall size of the BDD. In fact, many practical Boolean functions can efficiently be represented using BDDs, and efficient manipulations and evaluations are possible.

Example 2.31. Figure 2.23a illustrates the Shannon decomposition from (2.10). A binary decision diagram for the 1-bit full adder Boolean function, described by the truth table outlined in Fig. 2.8a, is given in Fig. 2.23b.

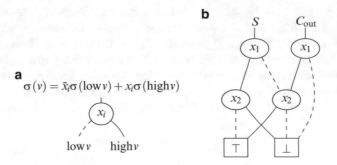

Fig. 2.23 Binary decision diagrams. (**a**) Shannon decomposition. (**b**) BDD for a 1-bit full adder

2.8.2 Quantum Multiple-Valued Decision Diagrams

A *quantum multiple-valued decision diagrams* (QMDD) [87] is a directed acyclic graph representing and manipulating $r^n \times r^n$ complex-valued matrices including the unitary matrices required to represent n-qubit quantum gates and circuits with r pure logic states. The fundamental idea is a recursive partitioning of the respective transformation matrix and the use of edge and vertex weights to represent various complex-valued matrix entries. More precisely, a transformation matrix of dimension $r^n \times r^n$ is successively partitioned into r^2 sub-matrices of dimension $r^{n-1} \times r^{n-1}$. This partitioning is represented by a directed acyclic graph (QMDD) with the following properties:

- There is a single terminal vertex representing the complex number 1 without any outgoing edge.
- Nonterminal vertices are labeled by an r^2 valued selection variable and have r^2 outgoing edges designated $e_1, e_2, \ldots, e_{r^2}$.
- There is a single root vertex which has a single incoming edge (the root edge) that itself has no source vertex.
- Every edge (including the root edge) has an associated complex-valued weight and edges with a weight of 0 (0-edges) point to the terminal vertex.
- There are no redundant vertices, i.e., no nonterminal vertex has r^2 identical outgoing edges (destinations and weights).
- Nonterminal vertices are unique, i.e., no two nonterminal vertices labeled by the same selection variable have the same set of outgoing edges (destinations and weights).
- Nonterminal vertices are normalized (see details in the following section).

Each assignment to the selection variables corresponds to choosing the respective sub-matrices in the partitioning process, and as a result, to some matrix entry. Thus, for any entry of the $r^n \times r^n$ matrix the QMDD can be evaluated in at most n steps by multiplying the weights on the path from the root vertex to the terminal vertex that is determined by the respective assignment.

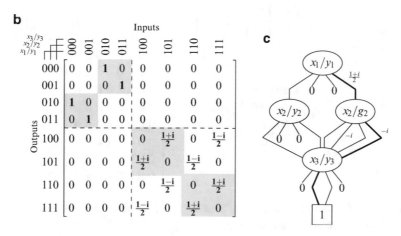

Fig. 2.24 Unitary matrix representations. (**a**) Quantum circuit based on the NCV library. (**b**) Transformation matrix. (**c**) QMDD

Example 2.32. Consider the NCV based circuit G of Fig. 2.24a. The transformation matrix, U_G, of the circuit is obtained by the multiplication of the unitary matrices of its gates in a reversed order.

The transformation matrix for the first gate, U_{g_1}, is the tensor product of the matrix representing a controlled V gate and a line which acts as an identity gate:

$$U_{g_1} = U_v \otimes U_I$$

$$= \begin{bmatrix} 1 & 0 & 0 & 0 \\ 0 & 1 & 0 & 0 \\ 0 & 0 & \frac{1+i}{2} & \frac{1-i}{2} \\ 0 & 0 & \frac{1-i}{2} & \frac{1+i}{2} \end{bmatrix} \otimes \begin{bmatrix} 1 & 0 \\ 0 & 1 \end{bmatrix}$$

$$= \begin{bmatrix} 1 & 0 & 0 & 0 & 0 & 0 & 0 & 0 \\ 0 & 1 & 0 & 0 & 0 & 0 & 0 & 0 \\ 0 & 0 & 1 & 0 & 0 & 0 & 0 & 0 \\ 0 & 0 & 0 & 1 & 0 & 0 & 0 & 0 \\ 0 & 0 & 0 & 0 & \frac{1+i}{2} & 0 & \frac{1-i}{2} & 0 \\ 0 & 0 & 0 & 0 & 0 & \frac{1+i}{2} & 0 & \frac{1-i}{2} \\ 0 & 0 & 0 & 0 & \frac{1-i}{2} & 0 & \frac{1+i}{2} & 0 \\ 0 & 0 & 0 & 0 & 0 & \frac{1-i}{2} & 0 & \frac{1+i}{2} \end{bmatrix}$$

The matrix of the NOT gate, U_{g2}, is calculated by the Kronecker product of the identity matrix, the NOT matrix, and the identity matrix:

$$U_{g2} = U_I \otimes U_I \otimes U_{NOT}$$

$$= \begin{bmatrix} 1 & 0 \\ 0 & 1 \end{bmatrix} \otimes \begin{bmatrix} 0 & 1 \\ 1 & 0 \end{bmatrix} \otimes \begin{bmatrix} 1 & 0 \\ 0 & 1 \end{bmatrix}$$

$$= \begin{bmatrix} 1 & 0 & 0 & 0 \\ 0 & 1 & 0 & 0 \\ 0 & 0 & 1 & 0 \\ 0 & 0 & 0 & 1 \end{bmatrix} \otimes \begin{bmatrix} 0 & 1 \\ 1 & 0 \end{bmatrix} = \begin{bmatrix} 1 & 0 & 0 & 0 & 0 & 0 & 0 & 0 \\ 0 & 1 & 0 & 0 & 0 & 0 & 0 & 0 \\ 0 & 0 & 0 & 1 & 0 & 0 & 0 & 0 \\ 0 & 0 & 1 & 0 & 0 & 0 & 0 & 0 \\ 0 & 0 & 0 & 0 & 0 & 1 & 0 & 0 \\ 0 & 0 & 0 & 0 & 1 & 0 & 0 & 0 \\ 0 & 0 & 0 & 0 & 0 & 0 & 0 & 1 \\ 0 & 0 & 0 & 0 & 0 & 0 & 1 & 0 \end{bmatrix}$$

The transformation matrix, U_G, that represents the circuit G is obtained by multiplying the matrices U_{g2} and U_{g1}:

$$U_G = \prod_{i=n}^{1} U_{gi} = U_{g2} \times U_{g1}$$

$$= \begin{bmatrix} 0 & 0 & 1 & 0 & 0 & 0 & 0 & 0 \\ 0 & 0 & 0 & 1 & 0 & 0 & 0 & 0 \\ 1 & 0 & 0 & 0 & 0 & 0 & 0 & 0 \\ 0 & 1 & 0 & 0 & 0 & 0 & 0 & 0 \\ 0 & 0 & 0 & 0 & 0 & 0 & 1 & 0 \\ 0 & 0 & 0 & 0 & 0 & 0 & 0 & 1 \\ 0 & 0 & 0 & 0 & 1 & 0 & 0 & 0 \\ 0 & 0 & 0 & 0 & 0 & 1 & 0 & 0 \end{bmatrix} \times \begin{bmatrix} 1 & 0 & 0 & 0 & 0 & 0 & 0 & 0 \\ 0 & 1 & 0 & 0 & 0 & 0 & 0 & 0 \\ 0 & 0 & 1 & 0 & 0 & 0 & 0 & 0 \\ 0 & 0 & 0 & 1 & 0 & 0 & 0 & 0 \\ 0 & 0 & 0 & 0 & \frac{1+i}{2} & 0 & \frac{1-i}{2} & 0 \\ 0 & 0 & 0 & 0 & 0 & \frac{1+i}{2} & 0 & \frac{1-i}{2} \\ 0 & 0 & 0 & 0 & \frac{1-i}{2} & 0 & \frac{1+i}{2} & 0 \\ 0 & 0 & 0 & 0 & 0 & \frac{1-i}{2} & 0 & \frac{1+i}{2} \end{bmatrix}$$

$$= \begin{bmatrix} 0 & 0 & 1 & 0 & 0 & 0 & 0 & 0 \\ 0 & 0 & 0 & 1 & 0 & 0 & 0 & 0 \\ 1 & 0 & 0 & 0 & 0 & 0 & 0 & 0 \\ 0 & 1 & 0 & 0 & 0 & 0 & 0 & 0 \\ 0 & 0 & 0 & 0 & 0 & \frac{1+i}{2} & 0 & \frac{1-i}{2} \\ 0 & 0 & 0 & 0 & \frac{1+i}{2} & 0 & \frac{1-i}{2} & 0 \\ 0 & 0 & 0 & 0 & 0 & \frac{1-i}{2} & 0 & \frac{1+i}{2} \\ 0 & 0 & 0 & 0 & \frac{1-i}{2} & 0 & \frac{1+i}{2} & 0 \end{bmatrix}$$

Hence, the unitary matrix describing the behavior of this circuit depicted in Fig. 2.24a is given in Fig. 2.24b. The QMDD for this matrix is given in Fig. 2.24c. The edges from each nonterminal vertex point to four sub-matrices. Each edge has a complex-valued weight. For clarity, edges with weights equal to 1 are omitted and edges with weight 0 are indicated as stubs. In fact, they point to the terminal vertex.

Here, the unique root vertex labeled by the pairing x_0/yo represents the whole matrix and has four outgoing edges targeting child vertices representing from left to right:

- the top-left sub-matrix where $x_1 = 0$ maps to $y_1 = 0$,
- the top-right sub-matrix where $x_1 = 1$ maps to $y_1 = 0$,
- the bottom-left sub-matrix where $x_1 = 0$ maps to $y_1 = 1$, and
- the bottom-right sub-matrix where $x_1 = 1$ maps to $y_1 = 1$.

This decomposition is repeated at each partitioning level until the terminal vertex that represents a single matrix entry $\boxed{1}$ is reached. The vertices are normalized by dividing the weights of all outgoing edges by a normalization factor (here: such that the leftmost edge with a nonzero weight has weight 1). This factor is propagated to referencing edges, e.g., the factor $\frac{1+i}{2}$ is propagated downwards from the x_1/y_1-level to the x_3/y_3-level in Fig. 2.24b. By this, structurally equivalent sub-matrices are compressed to a shared vertex (highlighted in same color in Fig. 2.24b). This procedure is repeated for each level until a single terminal vertex labeled by $\boxed{1}$ is created in the bottom. To obtain the value of a particular matrix entry, one has to follow the corresponding path from the root to the terminal vertex and multiply all edge weights on this path. For example, the matrix entry $\frac{1+i}{2}$ from the bottom "rightmost" sub-matrix of Fig. 2.24b can be determined as the product of the weights on the highlighted path of the QMDD in Fig. 2.24c.

A reversible function has the same number of inputs and outputs and maps each input pattern to a unique output pattern. In particular, a reversible function with n variables describes a permutation π_{2^n} of the set $\{0, 1, \ldots, 2^n - 1\}$. This permutation can also be described using a *permutation matrix*, i.e., a $2^n \times 2^n$ matrix Π_f as defined in Sect. 2.5.1. Since the QMDD [87] is a data structure that can represent the $r^n \times r^n$ unitary matrices used to describe quantum operations over r-valued logic. In the case where $r = 2$, a unitary matrix is in fact a permutation matrix. Hence, QMDDs are also suitable for representing and manipulating reversible functions.

Example 2.33. Figure 2.25a shows the truth table for the reversible function of a 1-bit full adder. Its permutation matrix is depicted in Fig. 2.25b. Its output pattern has only one input pattern. Therefore, each row has only one 1. The QMDD for the permutation matrix in Fig. 2.25b is shown in Fig. 2.25c. Two out of the sixteen 1-paths (paths from the root vertex to a $\boxed{1}$-terminal) have been randomly chosen and are highlighted in bold representing the mappings $0000 \mapsto 0000$ and $1110 \mapsto 1011$.

a

x_0	x_1	C_{in}	C	g_1	g_2	C_{out}	S
0	0	0	0	0	0	0	0
0	0	0	1	0	0	0	1
0	0	1	0	0	0	1	0
0	0	1	1	0	0	1	1
0	1	0	0	0	1	1	0
0	1	0	1	0	1	1	1
0	1	1	0	0	1	0	1
0	1	1	1	0	1	0	0
1	0	0	0	1	1	1	0
1	0	0	1	1	1	1	1
1	0	1	0	1	1	0	1
1	0	1	1	1	1	0	0
1	1	0	0	1	0	0	1
1	1	0	1	1	0	0	0
1	1	1	0	1	0	1	1
1	1	1	1	1	0	1	0

b

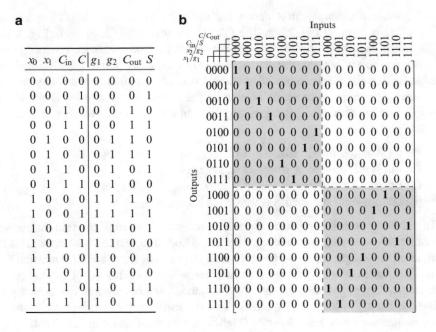

c

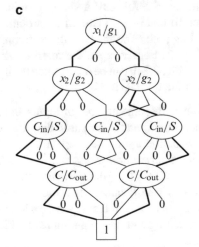

Fig. 2.25 1-bit full adder reversible function and its different representations. (**a**) Truth table. (**b**) Permutation matrix. (**c**) QMDD

Since QMDDs involve multiple edges from vertices and are applicable to both binary and multiple-valued problems, the QMDD package is not built using a standard decision diagram package, but the implementation employs well-known decision diagram techniques. QMDDs are ordered in a manner analogous to BDDs.

Hence, the vertex labels along any path from the start vertex adhere to a fixed ordering and no label appears more than once on any path. QMDDs are also reduced in a manner analogous to BDDs so each vertex represents a unique sub-matrix and the representation of common sub-matrices is shared.

The size of a unitary transformation matrix grows exponentially with respect to the number of input/output variables. However, QMDDs [87] provide an efficient data-structure which enables a much more compact representation of unitary matrices. QMDDs are canonical representations, if normalization of edge weights (as described in [87]) is performed. Thus, they are very convenient for equivalence checking.

Chapter 3
Optimizations and Complexity Analysis on the Reversible Level

Several synthesis approaches have been proposed to determine a reversible circuit realization for a given function. Significant improvements [46, 88, 132, 145] have been proposed. Yet, the majority of the synthesis approaches do not guarantee optimal realizations, in fact, the algorithms that do guarantee an optimal solution (e.g., [146]) are only applicable to small circuits of about 4–6 lines. Moreover, synthesis approaches are not aware of the technology for which the circuit is physically designed. As a result, several post-synthesis optimization approaches [88, 128] have been proposed to minimize a given circuit after it has been synthesized. Post-synthesis optimization approaches are used to reduce the circuits with respect to a given cost metric in quantum computing architectures, e.g., the number of lines, the depth, the cost to ensure linear nearest neighbor constraints, etc.

In this chapter, we aim at reducing the quantum cost and studying the complexity analysis of circuits in the reversible level. This chapter is structured as follows. Section 3.1 reviews the related work. Then, in Sects. 3.2 and 3.3, we give two approaches for the optimization of reversible circuits regarding the quantum cost metric. Section 3.4 describes a study for complexity analysis of reversible circuits and the chapter is concluded in Sect. 3.5.

3.1 Related Work

In the following, we review the different optimization algorithms. Then, we briefly summarize the existing work on the complexity of reversible circuits.

© Springer International Publishing Switzerland 2016
N. Abdessaied, R. Drechsler, *Reversible and Quantum Circuits*,
DOI 10.1007/978-3-319-31937-7_3

3.1.1   Optimization Approaches of Reversible Circuits

Due to the inherent complexity of reversible circuits, usually local optimization strategies are implemented, i.e., subcircuits are analyzed for possible reductions. The most employed optimization approaches are categorized into *rule-based optimization*, *template matching*, or *additional line based optimization*. These post-synthesis techniques attempt to apply templates or reduction rules by deleting identical gates or replacing circuits of gates with smaller ones using the moving rule [74, 78, 118].

3.1.1.1   Moving Rules

A *moving rule* enables the rearrangement of gates into a circuit without changing the functional behavior of a circuit. In the following, we review the classical moving rule as well as the different improvements proposed to extend this rule.

Classical Moving Rule

This rule was originally defined in [78] as the following property.

Property 3.1. Two adjacent gates can be interchanged if and only if the target for each gate is not a control for the other gate, i.e., in a reversible circuit, gate $G_1(C_1, t_1)$ can be interchanged with gate $G_2(C_2, t_2)$ if and only if $t_1 \notin C_2$ and $t_2 \notin C_1$.

Example 3.1. In the Fig. 3.1a, the first gate can be moved to the sixth position because its target does not coincide to any control of the gates 2, 3, 4, and 5. At the same time, its control line does not coincide with the targets of the gates 2, 3, 4, and 5. Meanwhile, the third gate cannot be moved to the fourth position because its target is in the same line as the control of the fourth gate.

Extended Moving Rule Based on LLP

As mentioned above, the classical moving rule is too restrictive and hence does not allow all possible gate reductions and deletions. Therefore, the classical moving rule was extended as described in [86, 110] allowing further moving abilities for each gate in the circuit. The new moving rule is defined in the following property.

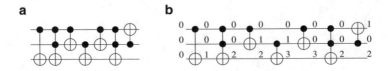

Fig. 3.1 Moving rules. (**a**) Classical moving rule. (**b**) Extended moving rule-based on LLP

Property 3.2. A gate can be moved from one end of a subcircuit to the other end if its controls are on lines that are *invariant* with respect to the circuit and its target is on a non-controlling line.

To identify whether a line is invariant, an algorithm called *line labeling procedure* (LLP) labels the line segments in a circuit [110]. This labeling is such that if two segments of a circuit line have the same label, those segments realize the same function. A label is an assignment of a number to each line after each gate. The labels are increased for targets which change the functionality, and are restored to a prior value when a set of gates realizing the identity operation is found [110]. If the label is the same at the beginning and the end of the segment, then the line is invariant.

Example 3.2. Figure 3.1b shows the result of applying LLP on a reversible circuit. The annotations in black outline the lines labeling after each gate. The sixth gate can be moved to the third position since its control lines have the same value in position 3 and 6 and therefore they are invariant. Also, its target line does not contain any control between the third and the sixth gate. This movement is not possible with the classical moving rule because the gate at position 6 is blocked by the fifth gate.

3.1.1.2   Rule-Based Procedure

Rule-based optimization (cf. [12, 33, 128, 130]) suggests a specific set of subcircuits together with cheaper replacements. Rule-based approaches are typically motivated by some synthesis techniques and often exploit reoccurring circuit structures. These approaches are not complete, i.e., they cannot guarantee an optimal circuit after optimization. However, the approaches are greedy meaning that optimization is always applied if a reduction can be achieved. Consequently, they suffer from not being able to escape local optima.

The degree of optimization in a rule-based approach depends on the considered subcircuit-rules (see the example of a subcircuits set outlined in Fig. 3.2) that are used in the search method to find matching subcircuits. The latter task is the most complex part in these algorithms.

Example 3.3. Figure 3.3 presents an optimization of a reversible circuit using the rule-based approach. The first circuit depicts the original circuit. The gates in positions 2 and 1 represent the first circuit of the rule given in Fig. 3.2a. Therefore

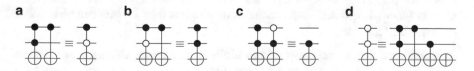

Fig. 3.2 Examples of MPMCT rules. (**a**) Rule 1. (**b**) Rule 2. (**c**) Rule 3. (**d**) Rule 4

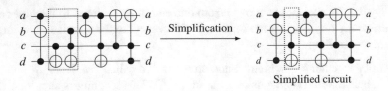

Fig. 3.3 Rule-based optimization

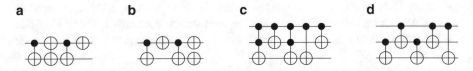

Fig. 3.4 Templates with 2 or 3 inputs from [79]. (**a**) Template 1. (**b**) Template 2. (**c**) Template 3. (**d**) Template 4

this circuit can be replaced by the second circuit which is cheaper in term of quantum cost (NCV-cost or T-depth). Hence the resulting circuit has a lower quantum cost than the original realization.

3.1.1.3 Template Matching Procedure

The template matching algorithm explained in [79, 88] is an efficient optimization technique. Given a set of templates which is a special class of identity circuits, the algorithm tries to determine subcircuits that match a part of a template. In this case, the determined subcircuit can be replaced with the inverted remaining part of the template due to reversibility. If the remaining part is smaller, the overall circuit size can be reduced.

Example 3.4. Figure 3.4 presents a set of templates with two or three inputs taken from [79].

Template matching approaches [78, 79, 119] are more powerful than rule-based approaches. A template is a generic circuit that realizes the identity function. Generic means that *one* template *represents* a *set* of identity circuits. Templates were initially proposed in [88] and their definition has been adjusted several times in the past. The current widely accepted definition [105] is as follows.

Definition 3.1 (Template). A template is a circuit with m gates that realizes the identity function such that each subcircuit of size less than $\lceil \frac{m}{2} \rceil$ cannot be reduced by any other template.

Note that a template consists of two different line types which we call C-lines and T-lines that should not be mistaken for control lines and target lines of a gate (cf. Fig. 3.5a). A C-line of a template is a line in which all gates only have

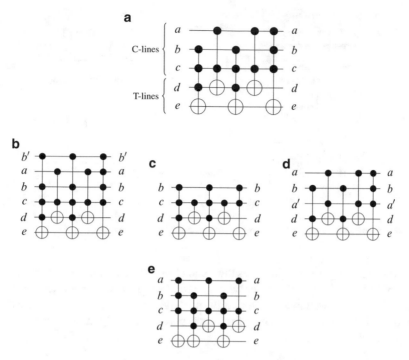

Fig. 3.5 Template disposition. (**a**) Original template. (**b**) Duplication. (**c**) Removal. (**d**) Replacement. (**e**) Rotation

control lines but no target lines, all other lines are T-lines. This separation is of use since any C-line can be duplicated, removed, or replaced without changing the functionality of the circuit. This is illustrated by means of Fig. 3.5b, c, and d, respectively. In addition, the order of the gates can be rotated being wrapped at the circuit boundaries, again without changing the functionality as it is an identity circuit. This is illustrated in Fig. 3.5e.

Since a template is an identity circuit it can be split anywhere in the middle and the left part equals the inverse of the right part and vice versa. The template matching algorithm is applied in order to reduce the circuit size or its costs. It takes a circuit and tries to find subcircuits that match a part of a template. If that part is longer or more expensive than the remaining part, the subcircuit is replaced by the inverse of that part. The matching procedure tries to find the first gate and then subsequently looks for the other gates which do not need to be necessarily adjacent by applying the time consuming moving rule. The search of subcircuits applicable for substitution is the bottleneck of the template matching algorithm for which several problem-solving methods have been proposed in the recent past [78, 79, 119]. Since these algorithms are based on heuristics, it cannot be ensured that a matching subcircuit can always be found.

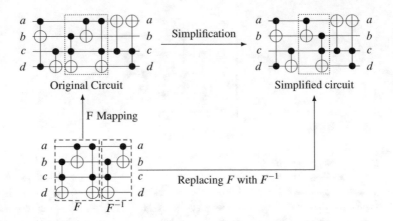

Fig. 3.6 Optimizing a reversible circuit based on template matching

Example 3.5. Figure 3.6 illustrates the basic template matching algorithm. It shows a template in the lower left corner with a left and a right part. A subcircuit matching the left part of the template is found in the original circuit; hence, this subcircuit can be replaced with the inverse of the right part of the template yielding a smaller simplified circuit of less cost.

3.1.1.4   Exploiting Additional Circuit Lines

Keeping the number of circuit lines as small as possible is important. This is mainly motivated by the fact that each circuit line has to be represented by a qubit, which is a very limited resource. Nevertheless, evaluations have shown that a (slight) extension of a circuit with additional lines may have significant benefits. For example in [15, 91, 98], it has been demonstrated that a larger number of circuit lines allow for a much cheaper mapping of reversible circuits to quantum circuits in terms of quantum cost.

In [145], evaluations showed that using twice the number of circuit lines reduces the quantum costs by up to two orders of magnitude. Eventually, this led to a post-synthesis optimization approach [33, 90, 152] which enables reductions in quantum costs of up to 69 % only by adding a single additional line to the circuit. The general idea of this optimization approach is to use new circuit lines in order to save factor of gates control lines, then the gates that have shared the same factor reuse the computed gate in the helper line. This reduces the size of these gates and thus decreases the quantum cost of the circuit since the controls of some gates are reduced.

Example 3.6. Figure 3.7 shows an example of an optimized reversible circuit based on an additional line. The original circuit is given in Fig. 3.7(a), The gates in this circuit have a common control factor. Hence, the quantum cost of this circuit can

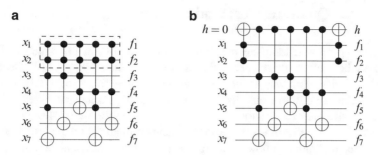

Fig. 3.7 Optimizing a reversible circuit with one additional line taken from [152]. (**a**) Original circuit. (**b**) With one additional line

be reduced as shown in Fig. 3.7(b) by adding an additional line h (at the top of the circuit), the factors are computed and saved on line h. After using the computed factor by the gates, the line h is reset to its original value. The application of this procedure leads to a reduction of the NCV-cost and T-depth from 190 and 114 to 70 and 42, respectively.

3.1.2   Complexity of Reversible Circuits

Motivated by the fact that upper bounds play a significant role in evaluating the complexity of synthesized reversible circuits, many methods for obtaining the upper bounds for given functions have been already studied.

In [76], it has been proven that every reversible function over n variables can be realized with less than

$$n \cdot 2^n \qquad (3.1)$$

MCT gates. This upper bound was obtained from transformation based synthesis approaches as initially presented in [88]. The algorithm traverses each of the 2^n rows of the truth table representation of a function and for each row adds at most one gate per column (i.e., variable).

In [77], it has been proven that every reversible function of 3 variables can be realized with 8 or less gates using the MCT library and 6 or less gates using the MPMCT gates. Then, in [52], it has been demonstrated that the implementation of any 4-bit reversible function requires at most 15 MCT gates.

3.2 Exact Quantum Cost Optimization

So far, the proposed template matching algorithms do not guarantee to find all possible subcircuits to that can be replaced by smaller ones [78, 79, 119]. Recently, a graph based algorithm was presented for exact template matching [105]. This algorithm can find all minimal circuits if a complete set of templates is given. However, it is applicable only for 3-qubit circuits. To overcome the search limitation of the template matching algorithm, we propose a new approach [1] that exploits Boolean satisfiability techniques allowing an exhaustive but yet efficient determination of subcircuits according to a given set of templates. For this purpose, the search for a subcircuit is formulated as a decision problem and encoded as a Boolean formula that is afterwards solved using an SMT solver. If the instance is satisfiable, the matching subcircuit can be replaced by the second half of the template. Otherwise, it can be concluded that the template cannot be used for further reductions of the circuit quantum cost. We have compared the exact template matching algorithm to the search method presented in [88, 119]. Experimental results show quantum cost reductions of up to 20 % in comparison with the circuit optimized by the method presented in [88, 119].

The remainder of the section is organized as follows. In Sect. 3.2.1 the general structure of our approach is explained, while details on the SMT encoding are provided in Sect. 3.2.2. Finally, experimental results are given in Sect. 3.2.3.

3.2.1 General Idea

In this section, we are proposing a new search method that determines matching subcircuits in a circuit for a given template. We do not change the general concepts for template matching as presented in [88]. Instead of applying heuristics for the search, we suggest an optimal and efficient template matching algorithm based on SMT which allows to exhaustively explore the full search space and therefore guarantees that a matched subcircuit is found if it exists.

Figure 3.8 outlines the proposed approach. Given a template with m gates and the original circuit, the proposed algorithm creates a Boolean formula encoding the decision problem whether there exists a subcircuit in the original circuit that matches the first k gates of the given template. While initially setting $k \leftarrow m$, k is decremented by 1 as long as the Boolean formula is unsatisfiable. If the formula is satisfiable the matched subcircuit can be extracted from the satisfying assignment that is returned from the SMT solver. The template length k is only decreased as long as the left part of the template is larger in cost than the right one.

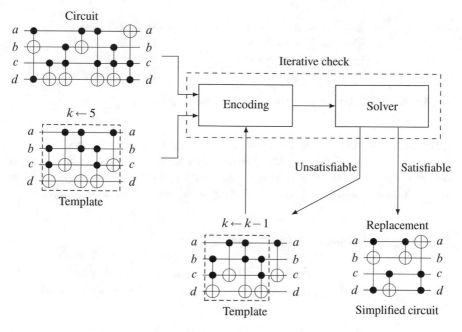

Fig. 3.8 Encoding the search method as an SMT instance

3.2.2 Encoding Using SMT

This section describes the proposed approach in detail. The encoding as a decision problem is explained by listing all constraints that are necessary for searching a correct match of a given template in a circuit.

3.2.2.1 Decision Problem

The overall decision problem can be stated as follows. Let $G = g_1 \cdots g_d$ be a circuit of n lines and $T = g'_1 \cdots g'_{d'}$ be a template of n' lines with $g_j = T_j(C_j, t_j)$ and $g_i = T_i(C_i, t_i)$ for $j = 1, \ldots, d$ and $i = 1, \ldots, d'$, respectively. The considered lines are defined over the variables $\{x_1, \ldots, x_n\}$ and $\{x'_1, \ldots, x'_{n'}\}$, respectively. Can the first $k \leq d'$ gates of the template T be matched in G, i.e., can positions $m_1, \ldots, m_k$ in G be found such that the gates can be moved together and match the first k gates of T.

Note that the order of lines in the template does not necessarily need to match the original order of lines in the circuit. As a result, besides the matching positions $m_1, \ldots, m_k$ also a line reordering $l_1, \ldots, l_{n'}$ is part of the solution, if the template can be matched such that the matched subcircuit must have lower quantum cost than the remaining subcircuit of the template.

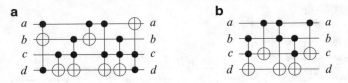

Fig. 3.9 Reversible circuit and template. (**a**) Circuit. (**b**) Template

Example 3.7. Figure 3.9a and b shows a circuit and a template, respectively. The first $k = 4$ gates of the template can be matched to the gates in the circuit at positions $m_1 = 3$, $m_2 = 4$, $m_3 = 5$, and $m_4 = 6$ when a line mapping $l_1 = 1$, $l_2 = 3$, $l_3 = 2$, and $l_4 = 4$, i.e., $x_1 \mapsto x'_1$, $x_2 \mapsto x'_3$, $x_3 \mapsto x'_2$, and $x_4 \mapsto x'_4$.

3.2.2.2 Gate Positions and Line Mapping

The variables $m_1, \ldots, m_k$ and $l_1, \ldots, l_{n'}$ represent gate and line indices in the intervals $[1, d]$ and $[1, n']$, respectively. A one-hot encoding is used for all these variables, i.e.,

$$\mathbf{m}_i \in \mathbb{B}^d \quad \text{for} \quad i = 1, \ldots, k$$

and

$$\mathbf{l}_j \in \mathbb{B}^{n'} \quad \text{for} \quad j = 1, \ldots, n'$$

such that $\nu \mathbf{m}_1 = \cdots = \nu \mathbf{m}_k = \nu \mathbf{l}_j = \cdots = \nu \mathbf{l}_{n'} = 1$, where the *sideways sum* ν denotes the number of all 1 bits in a bit-vector. The single bit that is set in the vectors denotes the respective index.

Two constraints are sufficient to enforce both the ordering of the positions m_i and guarantee the one-hot encoding, i.e.,

$$0 < \mathbf{m}_i < \mathbf{m}_{i+1} \quad \text{for} \quad i = 1, \ldots, k-1$$

and

$$\bigwedge_{i \neq j} (\mathbf{m}_i \wedge \mathbf{m}_j = 0) \wedge (\nu \mathbf{M} = k)$$

with $\mathbf{M} = \bigvee_{i=1}^{k} \mathbf{m}_i$ being the bit-mask containing all position indices. The line mapping variables do not have to follow an order; however all bit-vectors need to be one-hot encoded and they must all be different from each other, i.e.,

$$\bigwedge_{i \neq j} (\mathbf{l}_i \wedge \mathbf{l}_j = 0) \wedge \bigwedge_{j=1}^{n} \mathbf{l}_j \neq 0 \wedge \bigwedge_{j=1}^{n} \mathbf{l}_j = \underbrace{1 \ldots 1}_{n \text{ times}} .$$

3.2.2.3 Circuits and Templates

In order to represent circuits, we are adopting two different encodings which are both used later for restraining the mapping. They both share the property that control lines and targets are separately represented as bit-masks, however once in vertical and once in horizontal orientation. For the circuit's lines, we are following the horizontal scheme, i.e., for each control line and for each target line we add bit-vectors

$$\mathbf{c}_i \in \mathbb{B}^d \quad \text{with} \quad \mathbf{c}_i[j] \Leftrightarrow x_i \in c_j$$

and

$$\mathbf{t}_i \in \mathbb{B}^d \quad \text{with} \quad \mathbf{t}_i[j] \Leftrightarrow x_i = t_j$$

for $i = 1, \ldots, n$ and $j = 1, \ldots, d$. Note that the bit-vector indices start from 1.

Example 3.8. The encoded bit-vectors for the circuit lines of the initial circuit from Fig. 3.9a are illustrated in Fig. 3.10.

For the templates we are using a slightly different encoding that captures the different modification possibilities that exist for the templates and have been illustrated in Fig. 3.5. Furthermore, for the encoding we assume that the template can be extended to at most n lines such that it fits the circuit. Also, let us assume that the template has τ T-lines. Then the first $n - \tau$ circuit lines for the template are defined as follows.

$$\mathbf{c_i}' \in \mathbb{B}^k \quad \text{with} \quad \mathbf{c_i}' = \bigvee_{j=0}^{n'-\tau} \tilde{c}_j \qquad (i \leq n - \tau)$$

and

$$\mathbf{t_i}' \in \mathbb{B}^k \quad \text{with} \quad \mathbf{t_i}' = 0 \ldots 0 \qquad (i \leq n - \tau)$$

with

$$\tilde{c}_j \in \mathbb{B}^k \quad \text{with} \quad \begin{cases} 0 \cdots 0 & \text{if } j = 0, \\ \tilde{c}_j[\ell] \Leftrightarrow x_j' \in c_\ell & \text{otherwise.} \end{cases}$$

Fig. 3.10 Circuit encoding

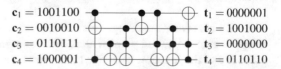

$$\begin{aligned}
c_1 &= 1001100 & t_1 &= 0000001 \\
c_2 &= 0010010 & t_2 &= 1001000 \\
c_3 &= 0110111 & t_3 &= 0000000 \\
c_4 &= 1000001 & t_4 &= 0110110
\end{aligned}$$

Fig. 3.11 Template encoding

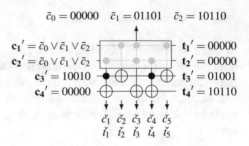

$$\tilde{c}_0 = 00000 \quad \tilde{c}_1 = 01101 \quad \tilde{c}_2 = 10110$$

$$\mathbf{c_1}' = \tilde{c}_0 \vee \tilde{c}_1 \vee \tilde{c}_2 \qquad \mathbf{t_1}' = 00000$$
$$\mathbf{c_2}' = \tilde{c}_0 \vee \tilde{c}_1 \vee \tilde{c}_2 \qquad \mathbf{t_2}' = 00000$$
$$\mathbf{c_3}' = 10010 \qquad \mathbf{t_3}' = 01001$$
$$\mathbf{c_4}' = 00000 \qquad \mathbf{t_4}' = 10110$$

$$\check{c}_1 \; \check{c}_2 \; \check{c}_3 \; \check{c}_4 \; \check{c}_5$$
$$\check{t}_1 \; \check{t}_2 \; \check{t}_3 \; \check{t}_4 \; \check{t}_5$$

That is, there are no targets on these lines and a control line can be either one of all possible C-lines (encoded as $\tilde{c}_{>0}$) or also $0 \ldots 0$ (encoded as $\tilde{c}_0$) indicates that the line is not used by the template. The lines for $i > n - \tau$ are encoded exactly in the same manner as for the original circuit lines:

$$\mathbf{c_i}' \in \mathbb{B}^k \quad \text{with} \quad \mathbf{c}_i[j] \Leftrightarrow x_i \in c_j \qquad (i < n - \tau)$$

and

$$\mathbf{t_i}' \in \mathbb{B}^k \quad \text{with} \quad \mathbf{t}_i[j] \Leftrightarrow x_i = t_j \qquad (i < n - \tau)$$

Example 3.9. The encoded bit-vectors for the circuit lines of the template from Fig. 3.9b are illustrated in Fig. 3.11. Notice that the order of the C-lines given in Fig. 3.9b does not matter anymore and is encapsulated in the $\tilde{c}_j$ variables.

From the bit-vectors $\mathbf{c_i}'$ and $\mathbf{t_i}'$ we are deriving new bit-masks $\check{c}_j$ and $\check{t}_j$ that describe the circuit in a vertical orientation, i.e., each bit-vector corresponds to a gate instead of a line. These are bit-vectors

$$\check{c}_j \in \mathbb{B}^n \quad \text{with} \quad \check{c}_j'[i] = \mathbf{c_i}'[j]$$

and

$$\check{t}_j \in \mathbb{B}^n \quad \text{with} \quad \check{t}_j'[i] = \mathbf{t_i}'[j].$$

Example 3.10. The encoded bit-vectors $\check{c}_j$ and $\check{t}_j$ that represent the controls and the target of each gate in the template are given in Fig. 3.11 at the bottom of each gate.

3.2.2.4 Mapping Template Gates to Circuit Gates

Given all bit-vector encodings from above, we can define the constraints that map template gates to circuit gates with respect to a line mapping. For this purpose, we make use of the function @ defined as follows.

$$@ : \mathbb{B}^n \times \mathbb{B}^n \qquad\qquad \rightarrow \mathbb{B} \qquad\qquad (3.6)$$

$$@ : (\mathbf{a}, \mathbf{b}) \qquad\qquad \mapsto \mathbf{a} \wedge \mathbf{b} \neq 0$$

where $\mathbf{a}, \mathbf{b} \in \mathbb{B}^n$. We use the function only in cases where $\mathbf{b}$ is one-hot encoded, hence, the function evaluates to true if and only if the one bit that is set in $\mathbf{b}$ is also set in $\mathbf{a}$. Given that, we can formalize the most important mapping for the encoding which maps the template gates to the circuit gates, expressed as

$$\check{c}_j @ \mathbf{l}_i = \mathbf{c}_i @ \mathbf{m}_j \quad \text{and} \quad \check{t}_j @ \mathbf{l}_i = \mathbf{c}_i @ \mathbf{m}_j$$

with $i = 1, \ldots, n$ and $j = 1, \ldots, k$. This formula is explained in the following. Assuming there is a control line in the jth gate of the template at the line where line i maps to. Then, there must be also a control line in the jth gate chosen by the mapping $\mathbf{m}_j$ in the original circuit at line i and vice versa. The same applies for target lines. Notice that the fact that we have vertical and horizontal encodings for the template and the circuit gates plays a key role in this encoding.

3.2.2.5 Moving Rule

The encoding given so far is not sufficient yet, since at the current state, arbitrary gates in the original circuits can be matched although it might not be possible to move them together. As a consequence, the moving rule needs to be encoded into the SMT instance as well.

In order to have a formal representation of the moving rule, we make use of the moving rule graph that has been introduced in [118]. In a moving rule graph, each vertex represents a gate of the circuit and two gates are connected by an edge if and only if the gates cannot be interchanged by moving.

Example 3.11. The moving rule graph for the circuit shown in Fig. 3.9a is depicted in Fig. 3.12.

We write $g_i < g_j$ if g_j cannot be moved before g_i and gate g_i cannot be moved past g_j. Since the moving rule is transitive, $g_i < g_j$ if and only if there exists a path between the vertex representing g_i and the vertex representing g_j in the moving rule graph.

Fig. 3.12 Moving rule graph for circuit in Fig. 3.9a

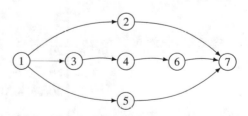

This leads to the final constraint which ensures the moving capabilities. For all $0 < i < j < k \leq d$ such that $g_i < \cdots < g_j < g_k$ and there is no j' such that $g_i < g_{j'} < g_k$, add the constraint

$$\mathbf{M}[i] \wedge \mathbf{M}[k] \Rightarrow \mathbf{M}[j].$$

These constraints can be easily generated with the help of the moving rule graph.

Example 3.12. Given the moving rule graph in Fig. 3.12, the extracted constraints are as follows.

$$\mathbf{M}[1] \wedge \mathbf{M}[7] \Rightarrow \mathbf{M}[2] \quad \mathbf{M}[1] \wedge \mathbf{M}[7] \Rightarrow \mathbf{M}[5]$$
$$\mathbf{M}[1] \wedge \mathbf{M}[7] \Rightarrow \mathbf{M}[6] \quad \mathbf{M}[1] \wedge \mathbf{M}[6] \Rightarrow \mathbf{M}[4]$$
$$\mathbf{M}[1] \wedge \mathbf{M}[4] \Rightarrow \mathbf{M}[3] \quad \mathbf{M}[3] \wedge \mathbf{M}[7] \Rightarrow \mathbf{M}[6]$$
$$\mathbf{M}[3] \wedge \mathbf{M}[6] \Rightarrow \mathbf{M}[4] \quad \mathbf{M}[4] \wedge \mathbf{M}[7] \Rightarrow \mathbf{M}[6]$$

3.2.2.6 Quantum Cost

With the SMT instance consisting of all the conditions encoded above, taking into account that $k \geq \left\lceil \frac{d'}{2} \right\rceil$, we obtain a gate-count reduction when a match is found. However as explained in Sect. 2.7.2, this optimization might not lead to a quantum cost reduction. Hence, it is important to mention that after finding a match, it is mandatory to check if the removal of the matched gates and replacing them by the second sequence of it may lead to a quantum cost reduction (the fact of replacing, adding, removing template control lines or rotating the template gates may affect the quantum cost of the mapped or the remain sequence of the template), otherwise the solution is not taken into account. In some cases, this can be repeated more than hundred times before finding a suitable match that brings a quantum cost reduction. Therefore, the quantum cost condition must be encoded into the SMT instance.

If the first k gates of a template match k gates in a circuit, then such subcircuit must be cheaper with respect to quantum cost to the reaming $d' - k$ gates of the templates. In order to determine the quantum cost of a given circuit, we define the variables $qc_1, \ldots, qc_k, q_1$, and, q_2 to represent the quantum cost of its gates, the quantum cost of the first k gates of a template, and the quantum cost of the remaining $d' - k$ gates, respectively.

The quantum cost [15] (qc_i) for each gate is calculated as follows.

$$qc_i = \begin{cases} 1 & \text{if } v\check{c}_i \leq 1, \\ 5 & \text{if } v\check{c}_i = 2, \\ 20 & \text{if } v\check{c}_i = 3, \\ 50 & \text{if } v\check{c}_i = 4, \\ 40(v\check{c}_i - 3) & \text{otherwise.} \end{cases}$$

for $i = 1, \ldots, d'$, where the *sideways sum* v denotes the number of all 1 bits in a bit-vector.

Note that the encoding above for the quantum cost is done with respect to the NCV-cost, while the quantum cost with respect to the T-depth is encoded as follows.

$$
qc_i = \begin{cases}
0 & \text{if } v\check{c}_i \leq 1, \\
3 & \text{if } v\check{c}_i = 2, \\
12 & \text{if } v\check{c}_i = 3, \\
30 & \text{if } v\check{c}_i = 4, \\
24(v\check{c}_i - 3) & \text{otherwise.}
\end{cases}
$$

The quantum cost of a subcircuit is defined as the sum of the quantum cost of its gates. This implies that the quantum cost of the first subcircuit of T with k gates is

$$
q_1 = \sum_{i=0}^{k} qc_i
$$

while the quantum cost of the remaining subcircuit with $d' - k$ is

$$
q_2 = \sum_{i=k+1}^{d'} qc_i.
$$

To make sure that the matched subcircuit in the circuit is cheaper than the remaining subcircuit, the following condition:

$$
q_1 > q_2
$$

must hold.

3.2.3 Experimental Results

The proposed approach has been implemented in C++. In order to read the reversible circuits as well as the templates, we used the open source toolkit for reversible and quantum circuits design *RevKit* [129]. The SMT problem, i.e., the template matching problem is encoded using the *metaSMT* [57] framework which provides the use of SAT and SMT solvers directly over its API. Different solvers can be used within the metaSMT framework; in our experiments the *Boolector* [23] SMT solver turns out to be the most efficient one.

We used specifications provided in [73, 147] as benchmarks. Each one has been embedded using the algorithm described in [136] and synthesized by the transformation based synthesis technique presented in [88]. We have adopted this synthesis algorithm that produces MCT circuits since the heuristic template

matching [88, 119] does not support circuits containing MPMCT gates when this
work is carried out. We have compared our approach to the one proposed in [88]
and used the same set of templates that is presented in the work.

Table 3.1 summarizes the obtained results for the conducted experiments. The
first five columns (Original benchmark (OB)) give the name of the circuits (ID) as
well as the number of lines (L), the number of gates (GC), the NCV-cost (NCV), and
the T-depth (TD). In the next columns, the obtained results for the heuristic template
matching approach (Heuristic TM (HTM)) and the proposed exact approach (Exact
TM (ETM)) are presented. For each circuit, the number of gates, the NCV-
cost (NCV), the T-depth (TD), and the run-times in seconds (Time) are given with
respect to the applied technique.

The later four columns (HTM/OB) present the quality regarding the NCV-cost
and the T-depths of the optimized circuits using the heuristic template matching
compared to the original circuits. The quality of optimized circuits using SMT
based template matching compared to the original circuits is given by the next four
columns (ETM/OB), and the improvements of the SMT based approach over the
heuristic approach are summarized by the last four columns (ETM/HTM).

Considering the quantum cost (NCV-cost or T-depth), for most of the circuits,
significant cost reduction can be observed when applying the new approach.
Applying the heuristic approach reduces the quantum costs by around 20 % for
all circuits. It can be clearly observed that these results are improved by the SMT
based approach. In fact, our proposed approach leads to an NCV-cost reductions
of 35% on average, 21 % in the worst case (*f2*), and 52 % in the best case (*4gt13*).
Similarly, our algorithm yields substantially cheaper circuits in terms of T-depth.
More precisely, an average improvement of around 37 % was accomplished by our
proposed algorithm. Furthermore, our technique achieved smaller NCV-cost and T-
depth for the majority of the benchmarks in comparison with the heuristic approach.
The additional improvement of the NCV-cost reaches 19 % on average, while the
T-depth can be further reduced with more than 19 %. The results clearly confirm
the impact of our approach to the circuit costs. The close similarity between the
reduction percentage of the NCV-cost and the T-depth returns to the fact that each
of them is a constant factor of the number of Toffolis in each MPMCT gate in a
circuit. The slight difference derives from considering the NOT and CNOT gates
with an NCV-cost of 1 and a T-depth 0.

3.2.3.1 Gate-Count Evaluation

The graph in Fig. 3.13 provides the number of gates for the original benchmarks (hashed-lines bar graph), the gate-count for the optimized benchmarks after
applying the heuristic approach (white bar graph), and the gate-count for the
optimized benchmarks via the SMT approach (dotted bar graph). We observe that
the circuits having the smallest quantum cost have the greatest gate-count in the

Table 3.1 Experimental results for heuristic and SMT based template matching

Original benchmark (OB)					Heuristic TM (HTM) [88]				Exact TM (ETM)				HTM/OB		ETM/OB		ETM/HTM	
ID	L	GC	NCV	TD	GC	NCV	TD	Time	GC	NCV	TD	Time	I_{NCV} (%)	I_{TD} (%)	I_{NCV} (%)	I_{TD} (%)	I_{NCV} (%)	I_{TD} (%)
sf	4	14	95	54	14	57	30	1.56	11	46	24	2.65	40.00	44.44	51.58	55.56	19.30	20.00
decod24	4	14	110	63	12	62	33	1.02	10	60	33	1.07	43.64	47.62	45.45	47.62	3.23	0.00
aj-e11	4	18	145	84	20	139	78	1.47	22	89	48	44.30	4.14	7.14	38.62	42.86	35.97	38.46
hwb4	4	19	157	90	20	124	69	1.59	24	84	45	10.50	21.02	23.33	46.50	50.00	32.26	34.78
4_49	4	23	215	126	22	138	78	1.64	21	111	63	41.90	35.81	38.10	48.37	50.00	19.57	19.23
rd32	4	27	242	141	29	236	135	1.68	30	169	93	31.30	2.48	4.26	30.17	34.04	28.39	31.11
4gt11	5	16	372	222	18	246	144	1.60	18	194	114	59.40	33.87	35.14	47.85	48.65	21.14	20.83
4gt13	5	18	476	285	20	319	186	1.62	19	225	132	89.60	32.98	34.74	52.73	53.68	29.47	29.03
4mod5	5	38	642	381	42	398	231	1.88	46	329	192	348.00	38.01	39.37	48.75	49.61	17.34	16.88
mod5d2	5	31	645	384	32	386	225	1.75	41	371	216	164.00	40.16	41.41	42.48	43.75	3.89	4.00
4gt10	5	26	666	399	30	602	357	1.72	34	502	297	186.00	9.61	10.53	24.62	25.56	16.61	16.81
alu	5	39	726	432	41	472	276	1.89	49	384	222	705.00	34.99	36.11	47.11	48.61	18.64	19.57
4mod7	5	41	803	477	45	770	456	1.87	61	497	291	764.00	4.11	4.40	38.11	38.99	35.45	36.18

(continued)

Table 3.1 (continued)

ID	Original benchmark (OB)				Heuristic TM (HTM) [88]				Exact TM (ETM)				HTM/OB		ETM/OB		ETM/HTM	
	L	GC	NCV	TD	GC	NCV	TD	Time	GC	NCV	TD	Time	I_{NCV} (%)	I_{TD} (%)	I_{NCV} (%)	I_{TD} (%)	I_{NCV} (%)	I_{TD} (%)
xor5	5	47	961	573	49	661	387	1.39	61	519	303	317	31.22	32.46	45.99	47.12	21.48	21.71
hwb5	5	58	1109	660	64	882	519	1.96	84	577	336	877.00	20.47	21.36	47.97	49.09	34.58	35.26
1-2-3	5	58	1245	744	62	1008	597	2.16	75	878	519	611.00	19.04	19.76	29.48	30.24	12.90	13.07
C17	6	117	3729	2232	127	3259	1944	2.76	178	2764	1647	6870.00	12.60	12.90	25.88	26.21	15.19	15.28
decod24-en.	6	115	3768	2256	125	3222	1920	2.79	168	2688	1602	28,600.00	14.49	14.89	28.66	28.99	16.57	16.56
ex3	6	110	3807	2280	118	3642	2175	3.25	167	2905	1734	4690.00	4.33	4.61	23.69	23.95	20.24	20.28
cm82a	6	121	4180	2502	130	3623	2163	3.61	181	3088	1836	3640.00	13.33	13.55	26.12	26.62	14.77	15.12
majority	6	124	4218	2526	134	3418	2040	3.77	186	2899	1731	8150.00	18.97	19.24	31.27	31.47	15.18	15.15
hwb6	6	134	4496	2691	144	3839	2292	4.06	207	3283	1956	12,300.00	14.61	14.83	26.98	27.31	14.48	14.66
ex2	6	136	4544	2721	146	4308	2577	4.08	210	3412	2037	4850.00	5.19	5.29	24.91	25.14	20.80	20.95
ham7	7	259	12994	7791	280	10,941	6546	10.00	392	9907	5928	20,900.00	15.80	15.98	23.76	23.91	9.45	9.44
rd53	7	290	13,332	7992	320	11,448	6852	12.40	432	9609	5751	27,900.00	14.13	14.26	27.93	28.04	16.06	16.07
f2	7	302	14,816	8883	328	13,208	7905	12.50	450	11,669	6990	25,500.00	10.85	11.01	21.24	21.31	11.65	11.57
hwb7	7	318	15,103	9054	345	13,842	8289	13.60	467	11,790	7059	25,700.00	8.35	8.45	21.94	22.03	14.82	14.84
Average													20.16	21.30	35.86	37.05	19.24	19.51

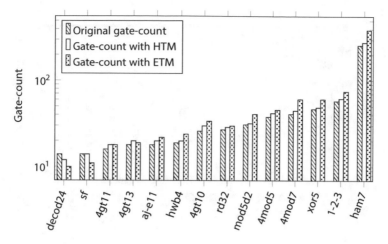

Fig. 3.13 Gate-count evaluation

majority of the cases. This clearly confirms that optimizing a reversible circuit with respect to the gate-count would not lead to an optimization of the quantum cost as explained in Sect. 2.7.2. As a hypothesis, circuits with larger number of reversible gates are likely to have smaller quantum cost.

3.2.3.2 Time Evaluation

The time evaluation is shown graphically in the plot in Fig. 3.14a. The values of the x-axis and the y-axis denote the benchmarks and the run-time, respectively. The plot contains two different scenarios: the run-time of the heuristic approach needed to optimize a circuit (white bar graph) and the run-time of our technique (dotted bar graph). Heuristic template matching is already known to be a time consuming process. However, one can clearly observe that the new approach needs an enormous computation time compared to the heuristic method. This is expected due to the fact that the match is determined exhaustively by the SAT solver, which needs higher run-time to provide the answer. Therefore a larger increase in the gate-count results in a sharply augmented run-time.

The plot in Fig. 3.14b outlines the required run time to optimize benchmarks with the same number of gates, but different numbers of lines. The time is correlated not only to the number of gates in a circuit but also to the number of lines because the considered set of templates is dynamically created with respect to the number of lines in the circuit. The larger number of lines the circuit has, the larger number of templates is considered in the search. Therefore, the required run time increases when the number of lines is increased.

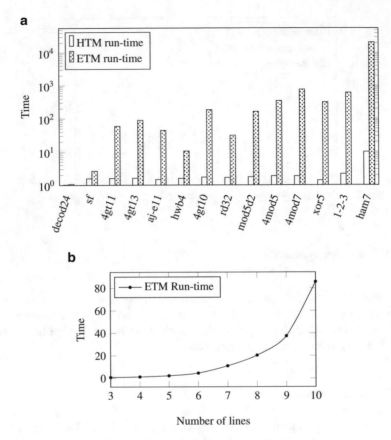

Fig. 3.14 Time evaluation. (**a**) Run time of the heuristic and exact template matching algorithms. (**b**) Correlation of the ETM runtime and the circuit lines

3.2.3.3 Synthesis Evaluation

Finally, to determine the best optimized realization with respect to NCV-cost and *T*-depth, the corresponding circuits for each benchmark are generated using the following MCT based synthesis approaches: the *transformation based synthesis* (TBS [88]) and the *Reed-Muller synthesis* (RMS [82]).

The experimental results are shown graphically in the plots in Fig. 3.15a. The values of the *x*-axis and the *y*-axis (logarithmic scale) denote the benchmark and the NCV-cost, respectively. Each plot contains two different scenarios including the NCV-cost of the circuits obtained from the RMS synthesis approach (white bar graph) and the NCV-cost of the circuits synthesized by the TBS synthesis approach (dotted bar graph). One can deduce that the circuits resulting from the Reed-Muller synthesis approach [82] in most of the cases have lower NCV-cost in comparison with the other TBS synthesis techniques. The similar observations are found for the *T*-depth as shown in Fig. 3.15b.

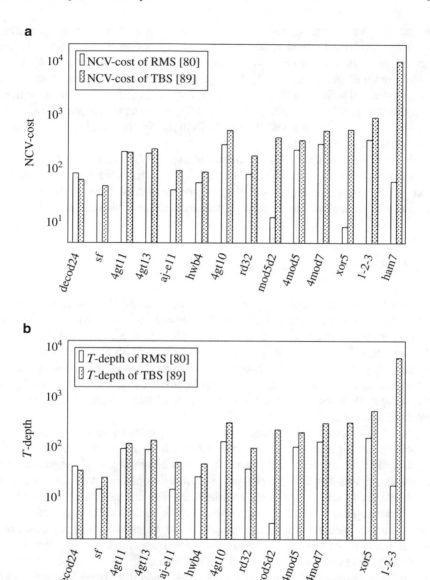

Fig. 3.15 Quantum cost evaluation of different synthesis approaches. (**a**) NCV-cost. (**b**) T-depth

3.3   Heuristic Quantum Cost Optimization

This section presents approaches to optimize the quantum cost of reversible circuits based on the rewrite rules from [127]. Two quantum cost optimization approaches are presented. The first method aims at reducing the cost by applying a greedy approach, whereas the second method is based on simulated annealing. Both improvements have been published in [6]. The application of simulated annealing can not only attain local optimum but also possibly global optimum [64], hence it can find further subcircuits to be replaced with smaller ones. The rewrite rules support this freedom by allowing a higher flexibility in gate movement compared to the existing moving rules. As confirmed by an experimental evaluation, improvements on quantum cost of up to 17 % in the first case and up to 30 % in the latter case can be observed.

The remainder of the section is structured as follows. The first section outlines the simulated annealing approach. Section 3.3.2 introduces the rewriting rules and Sect. 3.3.3 gives a detailed description of the implementation of the presented approach, and eventually the experimental results are evaluated in Sect. 3.3.4.

3.3.1 Simulated Annealing

Assume that S is a solution to a given problem. Let M be a *move* that can be performed on S. M will have an effect on the *cost* of S. The objective of optimization is to repeatedly perform *moves* to S such that the cost of S is reduced. The greedy algorithm is a simple heuristic optimization technique. With the problem at hand, it would only perform moves that lower the cost. The obvious problem with such an approach is that it can get stuck in a local optimum (see Fig. 3.16). Kirkpatrick et al. [64] suggest an optimization technique called *simulated annealing* that is based on statistical mechanics.

The idea of simulated annealing is based on random selection of moves (or transitions) on a given solution. Each move will have an effect on the cost of the solution. The moves can be cost-increasing, cost-decreasing, or cost-neutral. Cost-decreasing moves are always accepted. Simulated annealing uses the concept of *temperature* (it plays a significant role in statistical mechanics) to deal with the cost-increasing moves. Initially, the temperature is set to a quite large value and it decreases after each iteration by a cooling rate. Cost-increasing moves are accepted with a probability that depends on the temperature. In the beginning, many cost-increasing moves are accepted. As the temperature decreases, fewer and fewer moves are accepted. Finally, the procedure stops when the system reaches a stable state, i.e., no moves are accepted. The process of reversible circuit optimization is well suited for simulated annealing. Gates can be moved within the circuit. The move may be associated with an increased cost of the circuit. On the other hand, a gate may be moved adjacent to another gate that is identical. In this case both gates can be removed, since Toffoli gates are self-inverse, and the cost will be reduced.

Simulated annealing has been used in the synthesis of reversible circuits, e.g., in [109], the authors have presented a simulated annealing based Quine-McCluskey approach to synthesize a reversible circuit. Also, the synthesis algorithm presented in [32] has used simulated annealing to transform ESOP cubes.

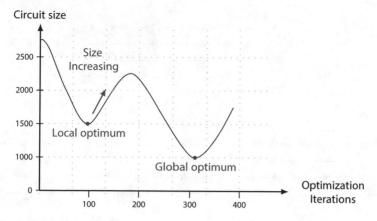

Fig. 3.16 Greedy heuristics via simulated annealing algorithm

Fig. 3.17 Rewriting rules.
(a) Rule 1. (b) Rule 2. (c)
Rule 3. (d) Rule 4

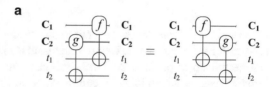

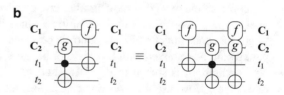

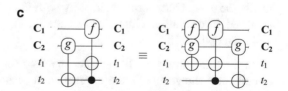

3.3.2 Rewriting Rules

The moving rules summarized in Sect. 3.1 are restraining the movement of gates into a circuit. On the other hand, the rewriting rules introduced in [127] are general for both MCT and MPMCT subcircuits and have more freedom for gate rearrangement. Based on these rules, we extract three scenarios for moving a gate from one position to another as shown in Fig. 3.17.

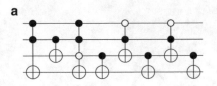

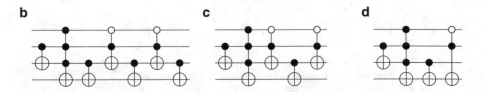

Fig. 3.18 Circuit optimization using different moving rules. (**a**) Original realization. (**b**) Classical moving rule. (**c**) Extended moving rule. (**d**) Rewriting rules

Example 3.13. Consider the reversible circuit depicted in Fig. 3.18a. Its NCV-cost is 39 and its T-depth is 21. Using the classical moving rule, the gate in position 1 can be moved to position 2. Hence, the gates in position 2 and 3 can be merged into one gate as shown in the circuit depicted in Fig. 3.18b. No more reductions are possible. The obtained circuit has an NCV-cost of 34 and a T-depth of 18. But if we consider the extended moving rule, one can move from the circuit in Fig. 3.18b also the gate in position 3 to the position 7 since its control line is invariant and its target line does not contain any controls between position 3 and 7. Thus the moved gate would be removed with its neighbor because they form an identity circuit. The obtained circuit is shown in Fig. 3.18c and has an NCV-cost of 32 and a T-depth of 18. Now if we use the rewriting rules, one can move the gate in position 3 from the circuit in Fig. 3.18c to position 4 by applying the rule shown in Fig. 3.17c. As a result, the moved gate forms an identity with the gate in position 5, and both can be removed. The resulting circuit is presented in Fig. 3.18d. It has an NCV-cost of 27 and a T-depth of 15.

In the following, we will consider the rewriting rules for optimizing reversible circuits and show through the experimental results its efficiency in reducing the quantum cost of reversible circuits.

3.3.3 Algorithms

To show the advantages of applying simulated annealing to reduce the quantum cost of reversible circuit, we compare it with the well known approach for optimizing reversible circuits based on exhaustive search. For this purpose, in this section we introduce as a first step a greedy approach combined with the rewriting rules. Then,

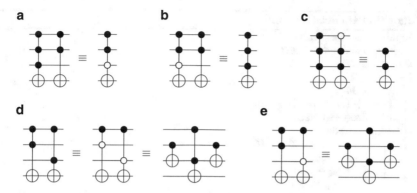

Fig. 3.19 Applied reductions. (**a**) Rule 1. (**b**) Rule 2. (**c**) Rule 3. (**d**) Rule 4. (**e**) Rule 5

as a second step, we define the simulated annealing approach. Both approaches are applied to subcircuits with a common target. Note that gates can be rearranged to create such a subcircuit using rewriting rules. Each MPMCT optimization procedure finds possible reductions in the circuit by moving gates across the circuit and making them adjacent. The gates may either be cancelled when they are identical or may be reduced to a less expensive subcircuit using the five different rules sketched in Fig. 3.19.

3.3.3.1   Greedy Approach

Given a reversible circuit $G = T_1(C_1, t_1) \ldots T_k(C_k, t_k)$ with k gates over variables $x_1, \ldots, x_n$, this algorithm optimizes the circuit by applying a greedy approach. For the computation, we are making use of the variables k, *pos*, and Δ_{cost} to denote the size of the circuit, the gate position, and the cost variable respectively. The remaining variables (i, j, and *optimized*) are used to control the algorithm loops.

The pseudo-code for the greedy algorithm is given in Algorithm 1. For each gate (line 2), this technique searches over the circuit for a gate that has the same target. A found gate (line 6) can be merged with the requested gate only when the rewriting process reduces the quantum cost of the targeted reversible circuit (line 10). When a reduction is applied, the optimization restarts the same process from the first gate of the circuit (line 18).

Algorithm 1: Greedy algorithm

Input: Reversible circuit G
Output: Optimized circuit G'

1   $k \leftarrow \texttt{Size}(G)$
2   $i \leftarrow 0$
3   **while** $i < k$ **do**
4      $j \leftarrow i + 1$
5   optimized $\leftarrow$ False
6   **while** $j < k$ **do**
7   pos $\leftarrow \texttt{SearchGate}(G, i, j + 1, k)$
8   **if** pos $< k$ **then**
9             $\Delta_{\text{cost}} \leftarrow \texttt{RewriteCost}(G, i, \text{pos})$
10            $j \leftarrow \text{pos} + 1$
11   **if** $\Delta_{cost} < 0$ **then**
12               $\texttt{RewriteMerge}(G, i, \text{pos})$
13   optimized $\leftarrow$ True
14   **end**
15   **else**
16            $j \leftarrow k$
17   **end**
18   **end**
19   **if** optimized **then**
20         $i \leftarrow 0$
21   **else**
22         $i \leftarrow i + 1$
23   **end**
24 **end**

3.3.3.2 Simulated Annealing Approach

Given a reversible circuit $G = T_1(C_1, t_1) \ldots T_k(C_k, t_k)$ with size k over variables $x_1, \ldots, x_n$. This algorithm optimizes the circuit by applying simulated annealing. For the computation, we are making use of the variables k, T, *frozen*, l, and Δ_{cost} to denote the size of the circuit, the used temperature, the stopping criterion, the number of generated perturbation, and the cost variable, respectively. The remaining variables (i, j, and *optimized*) are used to control the algorithm loops.

The algorithm is listed in Algorithm 2. We have chosen the initial temperature and the number of perturbation as factors of the number of gates in the initial circuit while the stopping criterion is set to 0 regardless of the size of the circuit and should not exceed 5 (see lines 2, 3, and 4). For a predetermined number of times l (line 9), the algorithm generates two different positions of gates (line 10 and 11) denoted by *loc* and *pos*. Then, it calculates the rewriting cost for rearranging them together (line 12). If the cost is decreased then the solution is accepted, i.e., the gates are merged together (line 14). Otherwise the solution is accepted with a certain probability (line 18). After each loop the temperature T is decreased (line 24) and the stopping criterion frozen is reset to 0 when the circuit has changed (line 26) otherwise the variable is incremented (line 28). This process is repeated until the stopping criterion reaches the value 5.

Algorithm 2: Simulated annealing

Input: Reversible circuit G
Output: Optimized circuit G'

1   $k \leftarrow \text{Size}(G)$
2   $T \leftarrow 10k$
3   frozen $\leftarrow 0$
4   $l \leftarrow 100k$
5   **while** frozen > 5 **do**
6     $j \leftarrow i + 1$
7   optimized $\leftarrow$ False
8     $i \leftarrow 0$
9   **for** $i = 0$ *to* l **do**
10   pos $\leftarrow \text{Random}(0,k)$
11   loc $\leftarrow \text{Random}(0,k)$
12       $\Delta_{\text{cost}} \leftarrow \text{RewriteCost}(G, \text{pos}, \text{loc})$
13   **if** $\Delta_{cost} < 0$ **then**
14         $\text{RewriteMerge}(G, \text{pos}, \text{loc})$
15   optimized $\leftarrow$ True
16   **else**
17         $q \leftarrow \text{Random}(0,1)$
18   **if** $q < e^{-\frac{\Delta_{cost}}{T}}$ **then**
19           $\text{RewriteMerge}(G, \text{pos}, \text{loc})$
20   optimized $\leftarrow$ True
21   **end**
22   **end**
23   **end**
24     $T \leftarrow 0.8T$
25   **if** optimized **then**
26   frozen $\leftarrow 0$
27   **else**
28   frozen $\leftarrow$ frozen $+ 1$
29   **end**
30 **end**

3.3.4 Experimental Results

The algorithms described above have been implemented in the open source toolkit *RevKit* [129]. The experimental evaluation has been carried out using the benchmarks taken from [73, 147], embedded by the procedure given in [136]. The obtained reversible functions are then synthesized by the QMDD based synthesis algorithm presented in [131] that scales well with the number of variables and generates circuits based on the MPMCT gate library. We have observed that our approaches lead to reversible circuits with smaller NCV-cost and T-depth compared to the approach presented in [86, 112].

The experimental results presented graphically depicted in Fig. 3.20a show the NCV-cost of optimized reversible circuits with respect to different optimization algorithms. The values of the x-axis and the y-axis denote the name of benchmarks and the NCV-cost, respectively. The plot contains three scenarios: the NCV-cost of

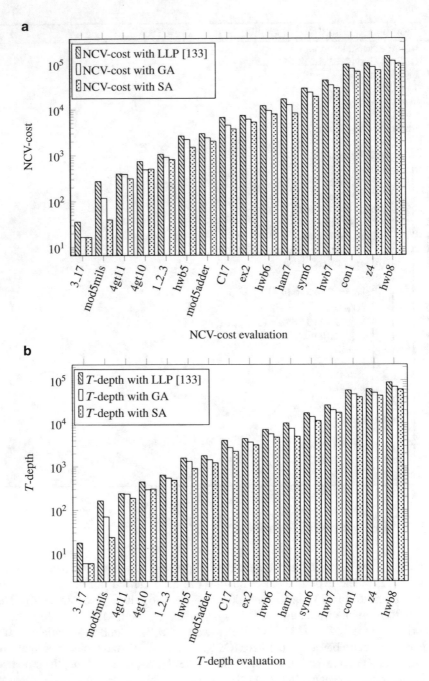

Fig. 3.20 Evaluation of optimization approaches regarding the quantum cost. (**a**) NCV-cost evaluation. (**b**) T-depth evaluation

optimized reversible circuits based on the LLP approach [130] (hashed-lines bar graph), the NCV-cost of circuits optimized with greedy approach (white bar graph), and the NCV-cost of the optimized circuits based on simulated annealing (dotted bar graph) One can clearly see that the greedy approach outperforms the LLP approach [130] based on the extended moving rule. While the simulated annealing approach produces circuits with the smallest NCV-cost in comparison with the other approaches. The same comparison is made with respect to T-depth in the plot shown in Fig. 3.20b.

Our results capture the following values: (1) the results of the line labeling procedure [130] (LLP), (2) the results of the greedy optimization approach (GA), (3) the results of optimized reversible circuits using the simulated annealing approach (SA), (4) the quality of optimized circuits using line labeling procedure compared to the original circuits (OB) that were synthesized using the synthesis approach described in [131], (5) improved quality with respect to the NCV-costs and the T-depths of the resulting circuits from the greedy approached compared to the original benchmarks, (6) the improvement of the greedy approach over the line labeling procedure, (7) the quality of optimized circuits using the simulated annealing compared to the original circuits, and (8) the improvement of the simulated annealing approach over the greedy approach.

The results are summarized in Tables 3.2 and 3.3. For each function, we show the number of lines (L), the NCV-cost (NCV), and the T-depth (TD). In addition to these metrics, we add the required run-time in seconds (Time) for each approach. The NCV-cost and the T-depth improvements are provided in the columns denoted by I_{NCV} and I_{TD}, respectively.

3.3.4.1 Greedy Approach

We have compared our greedy approach to the line labeling procedure. Table 3.2 summarizes the obtained results for the conducted experiments. We observed that the greedy approach outperforms the currently best known approach based on extended moving rule as outlined in Table 3.2.

For most of the functions, the greedy approach enables a reduction of T-depth and NCV-cost. The T-depth can be reduced by 21 % on average, 2 % in the worst case (*4gt11*), and 66 % in the best case (*3_17*), while the reduction of the NCV-cost with respect to the original benchmarks, reaches 21 % on average.

3.3.4.2 Simulated Annealing Approach

The results from the simulated annealing approach are given in third, fifth, and sixth columns of Table 3.3. Our proposed second approach leads to significant reductions on the T-depth and the NCV-cost. Over all circuits, reductions up to 34 % can be obtained on the NCV-cost. Also, it enables further improvements of the overall T-depth. The T-depth is reduced by 35 % on average, 15 % in the worst case (*pm1*), and in the best case (*mod5mils*) by 85 %.

Table 3.2 Experimental results for line labeling and greedy approaches

ID	Original benchmark (OB)				LL proc. (LLP)				Greedy ap. (GA)				LLP/OB		GA/OB		GA/LLP	
	L	GC	NCV	TD	GC	NCV	TD	Time	GC	NCV	TD	Time	I_{NCV} (%)	I_{TD} (%)	I_{NCV} (%)	I_{TD} (%)	I_{NCV} (%)	I_{TD} (%)
miller	3	6	26	15	6.00	26	15	0.44	5	17	9	0.00	0.00	0.00	34.62	40.00	34.62	40.00
3_17_6	3	7	37	18	7.00	37	18	0.06	5	17	6	0.00	0.00	0.00	54.05	66.67	54.05	66.67
hwb4	4	22	404	240	22.00	404	240	0.05	21	369	219	0.00	0.00	0.00	8.66	8.75	8.66	8.75
mod5mils	5	8	282	168	8.00	282	168	0.86	4	121	72	0.01	0.00	0.00	57.09	57.14	57.09	57.14
4gt11	5	13	410	246	13.00	410	246	0.04	12	401	240	0.01	0.00	0.00	2.20	2.44	2.20	2.44
4gt10	5	20	760	456	20.00	760	456	0.07	18	502	300	0.00	0.00	0.00	33.95	34.21	33.95	34.21
1_2_3	5	24	1093	654	24.00	1093	654	0.11	22	933	558	0.00	0.00	0.00	14.64	14.68	14.64	14.68
hwb5	5	61	2705	1617	61	2705	1617	0.05	55	2270	1356	0.08	0.00	0.00	16.08	16.14	16.08	16.14
mod5adder	6	42	3120	1872	42.00	3040	1824	0.12	36	2460	1476	0.04	2.56	2.56	21.15	21.15	19.08	19.08
C17	6	98	6970	4182	98.00	6830	4098	0.15	75	4635	2781	0.33	2.01	2.01	33.50	33.50	32.14	32.14
ex2	6	104	7660	4593	104.00	7528	4515	0.10	91	6230	3735	0.43	1.72	1.70	18.67	18.68	17.24	17.28
hwb6	6	161	12,340	7404	161.00	12,250	7350	0.12	136	9620	5772	0.97	0.73	0.73	22.04	22.04	21.47	21.47
ham7	7	172	17,160	10,296	172.00	17,160	10,296	0.36	145	12,840	7704	2.18	0.00	0.00	25.17	25.17	25.17	25.17
sym6	7	262	29,425	17,652	262.00	29,425	17,652	0.18	229	24,105	14,460	3.62	0.00	0.00	18.08	18.08	18.08	18.08

hwb7	7	382	44,405	26,643	382.00	44,405	26,643	0.18	321	34,605	20,763	14.06	0.00	0.00	22.07	22.07	22.07	22.07
con1	8	659	99,188	59,502	659.00	98,846	59,298	0.28	562	80,228	48,126	49.84	0.34	0.34	19.12	19.12	18.84	18.84
z4	8	693	105,300	63,171	693.00	104,878	62,919	0.28	600	86,918	52,143	80.80	0.40	0.40	17.46	17.46	17.12	17.13
hwb8	8	963	150,285	90,171	963.00	150,085	90,051	0.37	806	118,767	71,259	317.52	0.13	0.13	20.97	20.97	20.87	20.87
sqn	9	1238	237,979	142,779	1238.00	237,979	142,779	0.57	1048	192,739	115,635	803.10	0.00	0.00	19.01	19.01	19.01	19.01
rd73	9	1386	269,060	161,436	1386.00	269,060	161,436	0.77	1229	231,560	138,936	1274.75	0.00	0.00	13.94	13.94	13.94	13.94
sqrt8	9	1471	282,386	169,416	1471.00	282,386	169,416	1.40	1262	232,562	139,521	1029.56	0.00	0.00	17.64	17.65	17.64	17.65
radd	9	1513	292,305	175,374	1513.00	292,305	175,374	0.72	1304	241,651	144,978	1313.12	0.00	0.00	17.33	17.33	17.33	17.33
url1	9	1957	380,935	228,558	1957.00	380,935	228,558	0.81	1715	322,179	193,302	397.39	0.00	0.00	15.42	15.43	15.42	15.43
hwb9	9	2294	449,339	269,595	2294.00	449,339	269,595	1.40	2000	377,127	226,263	5096.60	0.00	0.00	16.07	16.07	16.07	16.07
root	10	2316	536,414	321,843	2315.00	535,734	321,435	1.53	1993	446,914	268,143	4875.54	0.13	0.13	16.68	16.69	16.58	16.58
max46	10	2784	652,100	391,260	2784.00	651,420	390,852	1.54	2383	539,584	323,748	8179.31	0.10	0.10	17.25	17.26	17.17	17.17
dist	10	2909	676,853	406,098	2909.00	676,013	405,594	1.50	2490	560,260	336,141	12,706.82	0.12	0.12	17.23	17.23	17.12	17.12
9symml	10	3848	901,974	541,167	3848.00	901,134	540,663	1.93	3329	756,818	454,071	40,639.88	0.09	0.09	16.09	16.09	16.01	16.02
sqr6	12	3725	1,157,282	694,356	3724.00	1,156,820	694,080	8.66	3227	979,286	587,556	4924.37	0.04	0.04	15.38	15.38	15.35	15.35
pm1	13	92	31,438	18,852	90.00	31,140	18,672	0.08	83	28,460	17,064	0.28	0.95	0.95	9.47	9.48	8.61	8.61
Average													0.31	0.31	21.03	21.66	20.79	21.41

Table 3.3 Experimental results for greedy and simulated annealing approaches

Original benchmark (OB)					Greedy ap. (GA)				Simulated an. (SA)				GA/OB		SA/OB		SA/GA	
ID	L	GC	NCV	TD	GC	NCV	TD	Time	GC	NCV	TD	Time	I_{NCV} (%)	I_{TD} (%)	I_{NCV} (%)	I_{TD} (%)	I_{NCV} (%)	I_{TD} (%)
miller	3	6	26	15	5	17	9	0.00	5	9	3	0.13	34.62	40.00	65.38	80.00	47.06	66.67
3_17_6	3	7	37	18	5	17	6	0.00	5	17	6	0.17	54.05	66.67	54.05	66.67	0.00	0.00
hwb4	4	22	404	240	21	369	219	0.00	19	233	135	0.63	8.66%	8.75	42.33	43.75	36.86	38.36
mod5mils	5	8	282	168	4	121	72	0.01	3	41	24	0.19	57.09	57.14	85.46	85.71	66.12	66.67
4gt11	5	13	410	246	12	401	240	0.01	11	321	192	0.38	2.20	2.44	21.71	21.95	19.95	20.00
4gt10	5	20	760	456	18	502	300	0.00	17	520	312	0.66	33.95	34.21	31.58	31.58	−3.59	−4.00
1_2_3	5	24	1093	654	22	933	558	0.00	22	835	498	0.83	14.64	14.68	23.60	23.85	10.50	10.75
hwb5	5	61	2705	1617	55	2270	1356	0.08	48	1550	924	3.99	16.08	16.14	42.70	42.86	31.72	31.86
mod5adder	6	42	3120	1872	36	2460	1476	0.04	34	2082	1248	2.45	21.15	21.15	33.27	33.33	15.37	15.45
C17	6	98	6970	4182	75	4635	2781	0.33	69	3817	2289	9.68	33.50	33.50	45.24	45.27	17.65	17.69
ex2	6	104	7660	4593	91	6230	3735	0.43	88	5404	3237	12.89	18.67	18.68	29.45	29.52	13.26	13.33
hwb6	6	161	12,340	7404	136	9620	5772	0.97	121	8090	4854	24.01	22.04	22.04	34.44	34.44	15.90	15.90
ham7	7	172	17,160	10,296	145	12,840	7704	2.18	124	8422	5046	31.98	25.17	25.17	50.92	50.99	34.41	34.50
sym6	7	262	29,425	17,652	229	24,105	14,460	3.62	209	19,460	11,667	77.23	18.08	18.08	33.87	33.91	19.27	19.32
hwb7	7	382	44,405	26,643	321	34,605	20,763	14.06	299	29,933	17,955	159.55	22.07	22.07	32.59	32.61	13.50	13.52

con1	8	659	99,188	59,502	562	80,228	48,126	49.84	514	68,854	41,298	574.35	19.12	19.12	30.58	30.59	14.18	14.19
z4	8	693	105,300	63,171	600	86,918	52,143	80.80	549	74,168	44,487	647.83	17.46	17.46	29.57	29.58	14.67	14.68
hwb8	8	963	150,285	90,171	806	118,767	71,259	317.52	739	103,415	62,043	1176.75	20.97	20.97	31.19	31.19	12.93	12.93
sqn	9	1238	237,979	142,779	1048	192,739	115,635	803.10	965	167,429	100,431	2675.11	19.01	19.01	29.65	29.66	13.13	13.15
rd73	9	1386	269,060	161,436	1229	231,560	138,936	1274.75	1131	200,550	120,312	3465.40	13.94	13.94	25.46	25.47	13.39	13.40
sqrt8	9	1471	282,386	169,416	1262	232,562	139,521	1029.56	1140	198,773	119,235	3562.76	17.64	17.65	29.61	29.62	14.53	14.54
radd	9	1513	292,305	175,374	1304	241,651	144,978	1313.12	1187	209,927	125,934	3903.36	17.33	17.33	28.18	28.19	13.13	13.14
urf1	9	1957	380,935	228,558	1715	322,179	193,302	397.39	1561	279,999	167,982	6394.80	15.42	15.43	26.50	26.50	13.09	13.10
hwb9	9	2294	449,339	269,595	2000	377,127	226,263	5096.60	1771	316,876	190,098	9185.13	16.07	16.07	29.48	29.49	15.98	15.98
root	10	2316	536,414	321,843	1993	446,914	268,143	4875.54	1835	392,250	235,323	16,103.30	16.68	16.69	26.88	26.88	12.23	12.24
max46	10	2784	652,100	391,260	2383	539,584	323,748	8179.31	2201	479,714	287,808	23,305.00	17.25	17.26	26.44	26.44	11.10	11.10
dist	10	2909	676,853	406,098	2490	560,260	336,141	12,706.82	2280	488,419	293,010	24,556.60	17.23	17.23	27.84	27.85	12.82	12.83
9symml	10	3848	901,974	541,167	3329	756,818	454,071	40,639.88	3033	662,461	397,434	47,124.00	16.09	16.09	26.55	26.56	12.47	12.47
sqrt6	12	3725	1,157,282	694,356	3227	979,286	587,556	4924.37	3017	894,670	536,772	62,393.70	15.38	15.38	22.69	22.69	8.64	8.64
pm1	13	92	31,438	18,852	83	28,460	17,064	0.28	79	26,700	16,008	14.91	9.47	9.48	15.07	15.09	6.18	6.19
Average													21.03	21.66	34.41	35.41	17.55	18.29

Table 3.4 Quantum cost variation

Original benchmark (OB)					Simulated annealing (SA)					
					NCV			TD		
ID	L	GC	NCV	TD	Min	Max	IV_{NCV} (%)	Min	Max	IV_{TD} (%)
3_17	3	7	37	18	17	17	0.00	6	6	0.00
mod5mils	5	8	282	168	41	132	105.20	24	78	105.88
4gt11	5	13	410	246	280	343	20.22	159	204	24.79
4gt10	5	20	760	456	415	582	33.50	246	348	34.34
1_2_3	5	24	1093	654	835	953	13.20	498	570	13.48
hwb5	5	61	2705	1617	1461	2370	47.45	870	1416	47.77
mod5adder	6	42	3120	1872	1534	2900	61.61	918	1740	61.85
C17	6	98	6970	4182	3507	4462	23.97	2103	2667	23.65
ex2	6	104	7660	4593	4462	7220	47.22	2667	4329	47.51
hwb6	6	161	12,340	7404	7277	11,035	41.04	4365	6621	41.07
ham7	7	172	17,160	10,296	7881	14,980	62.11	4719	8988	62.29
sym6	7	262	29,425	17,652	18,137	25,547	33.93	10,872	15,324	33.99
hwb7	7	382	44,405	26,643	28,681	38,325	28.79	17,199	22,995	28.84
con1	8	659	99,188	59,502	67,469	86,630	24.87	40,464	51,966	24.89
z4	8	693	105,300	63,171	71,886	93,320	25.95	43,119	55,983	25.96
hwb8	8	963	150,285	90,171	103,053	130,727	23.68	61,827	78,435	23.68
Average							37.05			37.50

As can be clearly observed, the effect of incorporating the simulated annealing algorithm for reversible circuit optimization is significant. By rearranging the gates in a random way, further reductions are achieved which confirms the idea outlined in Sect. 3.3.1. This approach outperforms the greedy approach for most of the functions. For example, the realization of the function *hwb4* is reduced by 8 % when the greedy approach is applied. Then, an additional 32 % of improvement is achieved by applying simulated annealing. In general, the simulated annealing approach leads to additional quantum cost reductions of 14 % on average compared to realizations optimized with the greedy approach.

Since the positions of gates to be rearranged are selected in a random way, different results for the same benchmark are obtained each time the simulated annealing is applied. Table 3.4 shows the NCV-cost and T-depth variation for each benchmark. For each benchmark, the simulated annealing is applied 100 time. Table 3.4 displays the lower (Min) and the upper (Max) extremes that an NCV-cost or a T-depth can take after applying the simulated annealing optimization. The percent difference is denoted by IV_{NCV} for the NCV-cost and IV_{TD} for the T-depth. Consider the benchmark *mod5mils* (originally it has an NCV-cost of 282), the optimization may results to an NCV-cost that varies between 42 and 132. The random choice of gates has a big impact on the reduction of the NCV-cost and the T-depth; the percent difference is 36 % and 37 % on average for the NCV-cost and the T-depth, respectively.

3.3.4.3 Time Evaluation

The graph in Fig. 3.21 indicates the gate-count on a reversible circuit on the x axis and the run-time in seconds on the y axis. The white bar graph shows the run-time of the greedy approach while the dotted bar graph depicts the run-time of the simulated annealing approach. The run-time of the greedy approach outperforms the run-time required by the simulated annealing. This is explained by the number of iterations needed by the latter approach to freeze the temperature.

3.3.4.4 Synthesis Evaluation

To determine the best realization synthesis approach with respect to the NCV-cost and T-depth, for each benchmark, the corresponding circuits based on the following two MPMCT based synthesis approaches and two MCT based synthesis approaches are generated: the QMDD based synthesis approach (QMDD) [131], the Young subgroups based synthesis approach (YSG) [135]), the Reed-Muller synthesis approach (RMS) [82]), and the transformation based synthesis approach (TBS) [88]). Then we optimized the obtained circuits using the simulated annealing approach.

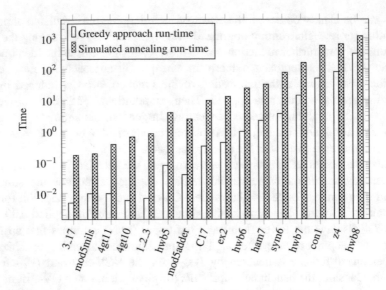

Fig. 3.21 Time evaluation

The experimental results are shown graphically in the plots of Fig. 3.22a. The values of the x-axis and the y-axis (logarithmic scale) denote the benchmark and the NCV-cost, respectively. Each plot contains four different scenarios: the NCV-cost of the optimized circuits generated by the QMDD synthesis (dashed-lines bar graph), the TBS synthesis (white bar graph), the YSG synthesis (dotted bar graph), and the RMS synthesis (black bar graph). In most of the cases, the optimized circuits originally synthesized by the RMS approach [82] outperforms the other synthesis techniques in terms of producing lower NCV-cost. The same observations are found for the T-depth as shown in Fig. 3.22b.

3.3.4.5   Exact Template Matching

In the following, we compare the template matching approach from Sect. 3.2 and the simulated annealing based approach. The experimental results are shown graphically in Fig. 3.23b. The values of the x-axis and the y-axis (logarithmic scale) denote the benchmark and the T-depth, respectively. Each plot contains three different scenarios: the T-depth of the original quantum circuits (dashed-lines bar graph), the T-depth of the optimized circuits based on simulated annealing (white bar graph), and the T-depth of the optimized circuits based on exact template matching (dotted bar graph). One can observe that the T-depth of the optimized circuits using exact template matching outperforms the T-depth obtained after applying simulated annealing. The same analysis are concluded for the NCV-cost as shown in Fig. 3.23a.

a

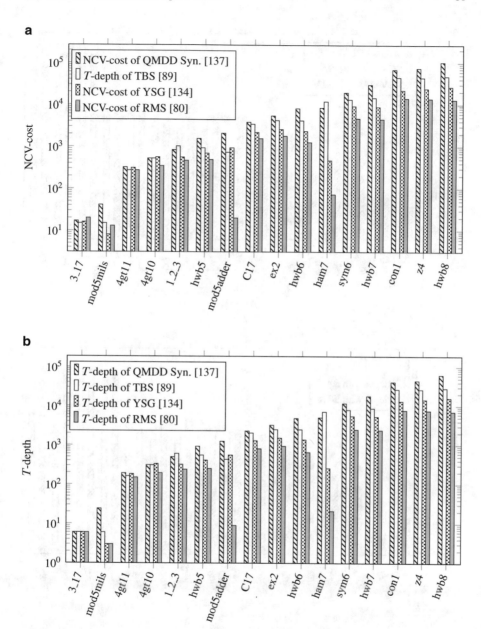

Fig. 3.22 Quantum cost of optimized circuits obtained from different synthesis approaches. (**a**) NCV-cost evaluation. (**b**) T-depth evaluation

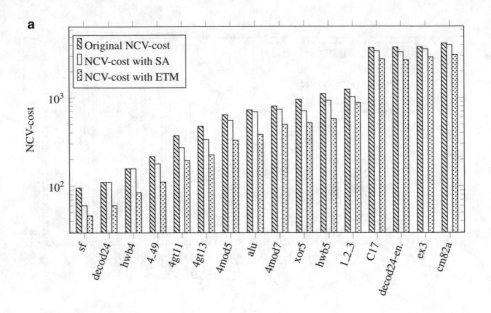

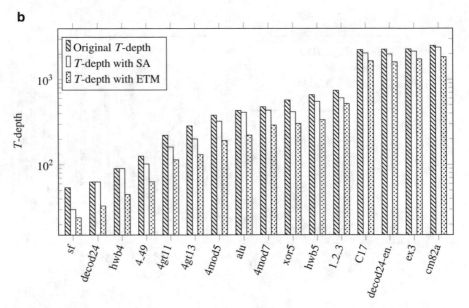

Fig. 3.23 Quantum cost resulting from SA and ETM. (**a**) NCV-cost evaluation. (**b**) *T*-depth evaluation

3.4 Complexity Analysis of Reversible Circuits

To compare the efficiency of different synthesis approaches, it is important to evaluate the resulting circuits. Depending on the target application, different metrics are applied to measure the complexity of a given circuit. Among these metrics, we find the number of gates, the gate delay (depth), and the quantum cost with respect to used quantum gate libraries. For more details the reader is referred to Sect. 2.7.

In this section, we investigate upper bounds on the number of gates that in general are needed to implement reversible circuits. This is important as a first approach to understand the complexity of reversible circuits, but, as mentioned, also to give an overall quality-measure of the different reversible synthesis methods. Previous research has investigated this topic based on specific synthesis algorithms and using a restricted gate library [76, 108]. The exact bounds are known for reversible circuits with at most four variables [52]. Furthermore, many studies have focused on the upper bound of the number of elementary quantum gates in reversible circuits [76, 108]. These upper bounds are specific to the special domain of quantum computing.

Our work is an improvement over previous reported upper bounds as well as an extension of the upper bound to a more general gate library. The improvement is achieved by combining a synthesis method based on Young subgroups [39] with decomposition of exclusive sum-of-products (ESOP) expressions into the different gate libraries [54, 149]. The work has been published in [4].

The following section describes in more depth the general idea, both of how to find the upper bounds and synthesizing using Young subgroups. In Sect. 3.4.3.1, the functions decomposition based upper bounds on MPMCT gates are described, while in Sect. 3.4.3.2, the ESOP expressions based upper bounds on MPMCT gates are explained and the improvements are discussed in Sect. 3.4.4.1.

3.4.1 Complexity of Single-Target Circuits

In this section we review the size of reversible circuits based on single-target gates

Given an n-variable reversible function $f(x_1, \ldots, x_n) \in \mathscr{B}_{n,n}$ We may decompose f into three n-variable reversible functions

$$f = g_1 \circ f' \circ g_2 \tag{3.7}$$

such that $g_1 = T_l(C, i)$ and $g_2 = T_r(C, i)$ are single-target gates with target line i and control functions l, r over the variable set $C = x \setminus x_i$, while f' is a function that does not change in line i; in other words $f'_i(x) = x_i$. Such a decomposition can always be found for all $1 \le i \le n$ [39].

A synthesis algorithm can readily be derived by recursively applying the decomposition to all variables. Eventually f' will be the identity function and the last two single-target gates g_1 and g_2 collapse to one gate $T_{l \oplus r}(X, i)$. This immediately implies a linear upper bound for single-target gates: each n-variable reversible function f can be realized with at most

$$uc_{ST}(n) = 2n - 1 \tag{3.8}$$

single-target gates [134]. If the decomposition is applied to all lines in numerical order, the resulting circuit is called a *V-shape*

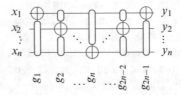

referring to the distribution of the target lines. A more detailed description of the algorithm can be found in [39].

The above synthesis approach, called the *Young subgroup synthesis*, was initially proposed in [39], then later re-implemented based on binary decisions diagrams (BDD) and presented in [135].

Example 3.14. Figure 3.24 shows an example application of the synthesis algorithm. First, the columns x_2, x_3, and f_2, f_3 are copied to the second and the fifth block of the extended truth table, respectively (the values are underlined). Afterwards, the columns x_1' and f_1' are filled in a way such that they have the same values columnwise by still ensuring reversibility. This is, e.g., done by filling x_1' with a 0 in the first row. To ensure reversibility in the fifth row a 1 is inserted since x_2 and x_3 have the same value. After that, the same value of x_1' is copied as a value to f_1' in the same row. Again, to ensure reversibility f_1' is assigned 0 in row 6. This process is repeated until all values of x_1' and f_1' have been assigned. Thereafter, the same step is performed for the second column on both sides of the extended truth table. Note that afterwards both middle blocks are equal except for the last column. After the extended truth table has been filled the control functions for the gates can be read. As an example, the first column in the second block changes whenever $\bar{x}_2 x_3$ or $x_2 \bar{x}_3$ holds, hence the control function is $x_2 \oplus x_3$.

3.4.2 Complexity of MPMCT Circuits

This section first gives upper bounds for the realization of single-target gates in terms of MPMCT gates and then gives upper bounds for reversible circuits in general.

To the best of our knowledge, no works thus far have studied the upper bound on the number of MPMCT gates in reversible circuits and compared them to upper

x_1 x_2 x_3	x_1' x_2 x_3	x_1' x_2' x_3	f_1' f_2' f_3	f_1' f_2 f_3	f_1 f_2 f_3
0 0 0	0 0 0	0 0 0	0 0 0	0 0 0	0 0 0
0 0 1	1 0 1	1 0 1	1 0 1	1 0 1	0 0 1
0 1 0	1 1 0	1 0 0	1 0 0	1 1 0	0 1 0
0 1 1	0 1 1	0 1 1	0 1 1	0 1 0	1 1 0
1 0 0	1 0 0	1 1 0	1 1 0	1 1 1	1 1 1
1 0 1	0 0 1	0 0 1	0 0 1	0 1 1	0 1 1
1 1 0	0 1 0	0 1 0	0 1 0	0 0 1	1 0 1
1 1 1	1 1 1	1 1 1	1 1 1	1 0 0	1 0 0

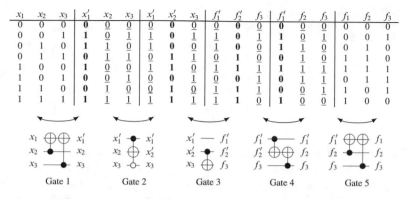

Fig. 3.24 Synthesis based on Young subgroups

bounds for MCT circuits. However, due to the incorporation of negative control lines in recent synthesis (e.g., [131]) and optimization approaches (e.g., [127]) this consideration becomes important.

3.4.3   Upper Bounds for Single-Target Gates

In the following sections, we describe different techniques to obtain $us(n)$ with respect to the MCT and MPMCT gate libraries.

3.4.3.1   Upper Bound Based on Function Decomposition

First, we present a method to obtain upper bounds based on exact synthesis techniques [54, 149] which guarantee a minimal circuit representation combined with function decomposition.

Theorem 3.1. *An n-bit single-target gate can be realized with at most*

$$us(n) = 3 \cdot 2^{n-3} - 2 \qquad (3.9)$$

MCT gates, if $n \geq 5$.

Proof. The proof is obtained by induction on n. For the base case let $n = 5$. Using exhaustive search we enumerated all 65,536 Boolean functions that can be represented by a 5-bit single-target gate and for each one we obtain the minimal circuit using an exact synthesis approach [54]. The largest circuit required $10 = 3 \cdot 2^{5-3} - 2$ MCT gates.

Fig. 3.25 Single-target gate decomposition. (**a**) MPMCT case. (**b**) MCT case

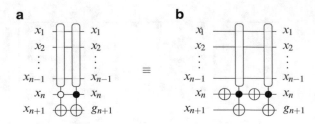

For the induction step, we assume that the result holds for n variables and consider a single-target gate using $n + 1$ lines. By applying the Shannon decomposition the control gate can be decomposed into two n-bit single-target gates and two NOT gates (see Fig. 3.25). According to the inductive hypothesis, the single-target gates in that circuit can be realized with at most $3 \cdot 2^{n-3} - 2$ gates. Hence, the single-target gate on $n + 1$ lines requires

$$2 \cdot (3 \cdot 2^{n-3} - 2) + 2 = 3 \cdot 2^{(n+1)-3} - 2$$

gates.   ∎

Theorem 3.2. *An n-bit single-target gate can be realized with at most*

$$us(n) = 3 \cdot 2^{n-4} \tag{3.10}$$

MPMCT gates, if $n \geq 5$.

Proof. We use the same proof as for Theorem 3.1. In the base case we used an exact synthesis approach that also considered negative control lines [149], and obtained $6 = 3 \cdot 2^{5-4}$ gates for the largest minimal circuit.

In the induction step we used the Shannon decomposition but the circuit construction without the additional two NOT gates (see Fig. 3.25) resulting in

$$2 \cdot (3 \cdot 2^{n-4}) = 3 \cdot 2^{(n+1)-4}$$

gates.   ∎

3.4.3.2 Upper Bound Based on ESOP Expressions

Tighter upper bounds for reversible circuits can also be obtained by combining the synthesis approach outlined in Sect. 3.4.1 using upper bounds for the size of ESOP expressions.

Table 3.5 Summary of upper bounds for realizing a single-target gate

Method	Single-target gate	
	MCT	MPMCT
Decomposition	$3 \cdot 2^{n-3} - 2$	$3 \cdot 2^{n-4}$
ESOP expressions	2^{n-1}	$29 \cdot 2^{n-8}$
Existing bound	$\frac{n2^n}{2n-1}$ [76]	N/A

Theorem 3.3. *An n-bit single-target gate can be realized with at most*

$$us(n) = 2^{n-1} \tag{3.11}$$

MCT gates.

Proof. This follows from the PPRM representation, which is canonical for a given function when disregarding the order of product terms. Hence, there exists a control function $g \in \mathcal{B}_{n-1}$ which PPRM expression consists of all 2^{n-1} product terms. ∎

Theorem 3.4. *An n-bit single-target gate can be realized with at most*

$$us(n) = 29 \cdot 2^{n-8} \tag{3.12}$$

MPMCT gates, if $n \geq 8$.

Proof. The best known upper bound on the number of product terms in a minimum ESOP form for an n-variables Boolean function is

$$29 \cdot 2^{n-7} \quad \text{with} \quad n \geq 7. \tag{3.13}$$

This upper bound was presented in [50]. Hence, the ESOP expression of the control function $g \in \mathcal{B}_{n-1}$ consists of at most $29 \cdot 2^{n-8}$ product terms. ∎

Table 3.5 summarizes the proposed upper bounds to represent a single-target gate using Toffoli gates in the MCT and MPMCT gate libraries.

3.4.4 Upper Bounds for Reversible Circuits

Using the complexity of a single-target gate (see previous section), a better upper bound is derived for MPMCT circuits.

Theorem 3.5. *Any reversible function over n variable can be implemented with at most, if $n \geq 8$*

$$uc_{\text{MPMCT}}(n) \leq 29(2n - 1) \cdot 2^{n-8} \tag{3.14}$$

MPMCT gates.

Table 3.6 Summary of upper bounds for representing a reversible function

Method	Reversible circuit	
	MCT	MPMCT
Decomposition	$3n2^{n-2} - 4n - 3 \cdot 2^{n-3} + 2$	$3n2^{n-3} - 3 \cdot 2^{n-4}$
ESOP expressions	$n2^n - 2^{n-1}$	$29n2^{n-7} - 29 \cdot 2^{n-8}$
Existing bound	$n2^n$ [76]	N/A

Table 3.7 Upper bounds on number of gates in a reversible circuit

Upper bound	Number of variables									Impr. (%)
	2	3	4	5	6	7	8	9	10	
Existing	8	24	64	160	384	896	2048	4608	10,240	
ESOP (MCT)	6	20	56	144	352	832	1920	4352	9728	11
Decomposition (MCT)	6	20	42	90	242	598	1410	3230	7258	31
Decomposition (MPMCT)	3	10	21	54	132	312	720	1632	3648	64
ESOP (MPMCT)	3	10	21	54	99	208	435	986	2204	71

Proof. Since each reversible function of n variables can be realized as a circuit consisting of $2n - 1$ single-target gates based on the Young subgroup synthesis approach, any MPMCT circuit can be synthesized in

$$(2n - 1) \cdot us(n) \tag{3.15}$$

where $us(n)$ is the largest number of Toffoli gates required to represent a single-target gate. The best upper bound for single-target gate is $29 \cdot 2^{n-8}$. Hence the upper bound of MPMCT circuits is $29(2n - 1) \cdot 2^{n-8}$. ∎

Based on the different upper bounds of a single-target gate, Table 3.6 summarizes the tighter upper bounds to represent a reversible circuit using Toffoli gates in the MCT and MPMCT gate libraries.

3.4.4.1 Analysis

To get a better intuition how the new bounds compare to the existing one, we listed their absolute values for up to 10 variables in Table 3.7. The first row gives the number of lines in the reversible circuits. In the next rows, the obtained results for the existing upper bound and the proposed upper bounds are presented. For the proposed approaches the average improvement compared to the existing upper bound is given in the last column.

The upper bound for the MCT gates obtained by function decomposition shows better improvements compared to the MCT gate upper bound derived using ESOP minimization. Exploiting the exact synthesis approach in the calculation of the MCT upper bound enables further improvements, which reach 31 % on average.

On the other hand, results clearly confirm that the ESOP minimization based upper bound on the number of MPMCT gates is better than the upper bound obtained by the approach using function decomposition.

By applying the ESOP based upper bound, an average improvement of 71 % can be obtained compared to the best known upper bound. Furthermore, there is an additional improvement of 31 % in comparison with the best MCT gates upper bound.

For the best upper bound, the overall improvement is

$$1 - \frac{(2n - 1) \cdot 29 \cdot 2^{n-8}}{n2^n} = 1 - \frac{29}{128} + \frac{29}{n2^8} > 1 - \frac{29}{128} \approx 77\,\%. \qquad (3.16)$$

The overall improvement can be calculated in the same way for the remaining bounds as well.

3.5 Summary

In this chapter, we gave a new search method for applying templates in the template matching algorithm. Instead of using heuristics for matching gates, the problem is encoded into an instance of Boolean satisfiability and therefore an exhaustive examination of the search space is ensured. The proposed approach ensures that these subcircuits are always found if they exist. Experimental results demonstrate that template matching yields smaller circuits when applying the new method for a subcircuit determination. The outlined approach leads to improvements of 19 % on average in terms of quantum cost compared to the heuristic template matching approach. However, this approach is feasible only for small circuits because the response time grows excessively fast according to the number of gates.

Then, we introduced optimization approaches that are scalable with large reversible circuits based on rewriting rules. We presented two different strategies: a greedy approach and a simulated annealing approach. The proposed algorithms achieve significant reductions (with respect to the NCV-cost and T-depth), specially when simulated annealing is considered. Experimental results show that the given approaches enable reductions of the quantum costs by up to 17 % on average compared to the currently known techniques.

Finally, we studied the complexity of reversible circuits and gave better upper bounds for MCT as well as MPMCT based reversible circuits compared to the known bounds given in [76]. Based on the Young subgroups synthesis approach, we demonstrated that a reversible function of n variables can be implemented with at most $3n2^{n-3} - 3 \cdot 2^{n-4}$ MPMCT gates.

Chapter 4
Optimization and Complexity Analysis on the Mapping Level

The common gate library for the synthesis of reversible circuits consists of mixed-polarity multiple-control Toffoli gates or single target gates. Such gates offer a convenient representation to model the functionality of a reversible circuit but are not universal for quantum operations. Many aspects, particularly those considering fault tolerance and error correction properties, cannot be considered effectively at this abstraction level. Consequently, after deriving and optimizing a reversible circuit for a given function as it is explained in the previous chapter, the next step consists of mapping the circuit into a quantum circuit. For this purpose, the following steps are usually applied: (1) Performing circuit transformation such that the circuit consists only of NCT gates. (2) Mapping each 2-control Toffoli gate to an optimum quantum circuit composed of gates from a given library. Many algorithms have been proposed to accomplish the first step, i.e., to map an MPMCT circuit to an NCT circuit. In the following section we review the different mapping strategies for MPMCT or ST based circuits. Then, in Sect. 4.2 we introduce an enhanced mapping approach for the ST based circuits. Later, in Sect. 4.3 we propose an efficient mapping methodology targeting the Clifford $+$ T library based circuits in order to minimize their T-depth. Finally, we study the complexity of mapped circuits, i.e., reversible circuits based on the NCT gate library in Sect. 4.4.

4.1 Related Work

In the following, we review the different mapping strategies used to decompose MCT and MPMCT circuits. Then, we briefly summarize the existing work on the complexity of NCT circuits.

© Springer International Publishing Switzerland 2016
N. Abdessaied, R. Drechsler, *Reversible and Quantum Circuits*,
DOI 10.1007/978-3-319-31937-7_4

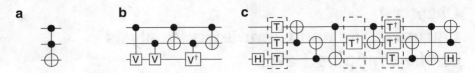

Fig. 4.1 Quantum mapping of a Toffoli gate. (**a**) Toffoli gate. (**b**) NCV circuit. (**c**) Clifford $+$ T circuit

Fig. 4.2 Mapping an MCT gate using the Barenco et al. (Lemma 7.2). (**a**) 4-control gate. (**b**) Barenco et al. (Lemma 7.2)

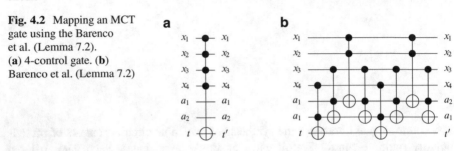

4.1.1 Mapping Approaches

Different mapping techniques have been introduced to transform an MPMCT gate into an NCT circuit. The complexity of the resulting circuit depends mainly on the number of used ancillas. Free ancilla mapping algorithms are either exponential [15, Lemma 7.1] or polynomial [138], whereas mapping approaches using additional ancillas produce, for a given MPMCT gate, a linear complexity NCT circuit [15, 91, 98]. In the following, we review the proposed mapping algorithms that transform an MPMCT gate into an NCT circuit consisting of a linear number of gates.

4.1.1.1 Barenco et al. Mapping

Barenco et al. Lemma 7.2 According to [15], a c-control MCT gate with $c - 2$ available ancillae can be mapped directly to an NCT circuit that consists of

$$4(c - 2) \tag{4.1}$$

2-control Toffoli gates. Since each Toffoli gate has an NCV-cost of 5 and a T-depth of 3 as shown in Fig. 4.1, the obtained NCT circuit has an NCV-cost of

$$20(c - 2) \tag{4.2}$$

and a T-depth of

$$12(c - 2). \tag{4.3}$$

Example 4.1. Figure 4.2b illustrates resulting circuit after mapping a 4-control MCT gate using Lemma 7.2 since it has two additional helper lines. The circuit comprises $4(4 - 2) = 8$ 2-control Toffoli gates. The resulting circuit has an NCV-cost of $5 \cdot 8 = 40$, and a T-depth of $3 \cdot 8 = 24$.

In [75], Maslov and Dueck have optimized the above algorithm to reduce the NCV-cost of NCV based circuits. Each Toffoli gate is replaced by a Peres gate

$$\tag{4.4}$$

or its inverse

$$\tag{4.5}$$

Both gates use four NCV gates instead of 5, as it is depicted above. The mapping algorithm explained above is then further optimized by removing the first V gate from each Peres gate and the last $V^\dagger$ gate from each inverse Peres gate using the moving and reduction rules defined in [83]. Therefore, each Peres or its inverse gate is replaced by the αP gate

$$\tag{4.6}$$

where the inverse of the αP gate is the same as the αP gate. Note that the two upper "points" of the circuit are left as Peres gates since they cannot be further optimized. To summarize, according to Maslov and Dueck, each Toffoli gate resulting from the mapping algorithm based on Lemma 7.2 is mapped to αP gates except two of them are mapped to Peres gates. Hence, the resulting circuit has an NCV-cost of

$$12(c - 2) + 2. \tag{4.7}$$

Example 4.2. The circuit depicted in Fig. 4.3b presents the original transformation for the 4-control gate with 2 ancillae while the circuit depicted in Fig. 4.3c is the mapped circuit with respect to the mapping such that each Toffoli gate is replaced

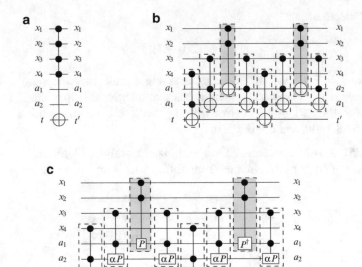

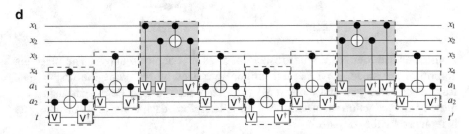

Fig. 4.3 NCV-cost optimized Barenco et al. (Lemma 7.2) mapping. (**a**) 4-control gate. (**b**) Barenco et al. (Lemma 7.2). (**c**) Toffoli gates replacement. (**d**) Optimized mapping regarding the NCV-cost

by either a Peres gate, its inverse (in grey rectangle) or an αP gate. The circuit has an NCV-cost of $12(4-2)+2=26$ instead of 40 when no optimization is applied.

This approach is also valid for MPMCT gates. In [83], the authors proved that an MPMCT gate with some but not all negative controls can be implemented with the same NCV-cost as an MCT gate with the same number of controls.

Barenco et al. Lemma 7.3 According to [15], an MPMCT gate $T(C, t)$ with $|C| \geq 3$ is mapped to the following 4 gates circuit:

$$T(C_1, a_1) \circ T(C_2 \cup \{a_1\}, t) \circ T(C_1, a_1) \circ T(C_2 \cup \{a_1\}, t) \qquad (4.8)$$

where $C = C_1 \cup C_2$, $C_1 \cap C_2 = \emptyset$, $|C_1| = \lceil \frac{|C|}{2} \rceil$, and $|C_2| = |C| - |C_1|$.

The resulting circuit consists of two identical gates with $|C_1|$ controls and two other identical gates with $|C_2|$ controls, such that each of them is placed alternately.

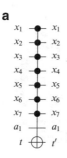
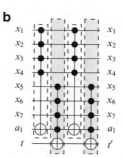

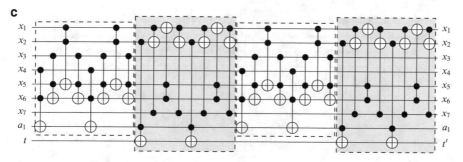

Fig. 4.4 Mapping an MCT gate using the Barenco et al. (Lemma 7.3). (**a**) 7-control gate. (**b**) Barenco et al. (Lemma 7.3). (**c**) Final mapping based on Barenco et al. mapping

If no free line is available, an additional line must be added to the circuit. Note that the circuit restores the value on a_1 and therefore it can be reused for all gates. The *ancilla line* a_1 can neither be in C nor can it be t.

Example 4.3. Figure 4.4a shows a 7-control MCT gate. Applying Lemma 7.3 mapping algorithm results in the circuit depicted in Fig. 4.4b. Note that a_1 is considered as the *helper line*, $|C_1| = \lceil \frac{7}{2} \rceil = 4$, and $|C_2| = 7 - |C_1| = 3$.

The Barenco et al. (Lemma 7.2) algorithm can be applied as a second step to map each of the four gates obtained from the Barenco et al. (Lemma 7.3) algorithm as explained above. The final circuit consists of

$$8(c - 3) \tag{4.9}$$

2-control Toffoli gates, an NCV-cost of

$$24(c - 3) + 8 \tag{4.10}$$

and a T-depth of

$$24(c - 3). \tag{4.11}$$

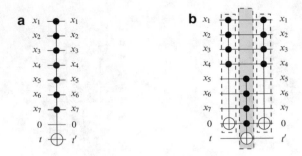

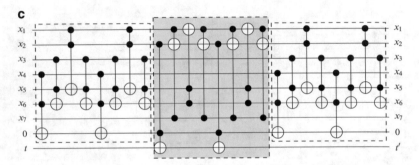

Fig. 4.5 Mapping an MCT gate using the Nielsen and Chuang mapping. (**a**) 7-control gate. (**b**) Nielsen and Chuang mapping. (**c**) Final mapping based on Nielsen and Chuang mapping

In the rest of this book, we denote the one ancilla mapping algorithms (Lemma 7.3) and (Lemma 7.2) proposed by Barenco et al. by B1 and B2, respectively.

Example 4.4. The four obtained gates after mapping the 7-control gate using the B1 mapping shown in Fig. 4.4b have enough ancillae to apply the B2 mapping. Therefore, each of them is mapped to a Toffoli circuit. The final circuit is given in Fig. 4.4c and it consists of $8(7 - 3) = 32$ Toffoli gates. The resulting circuit has an NCV-cost of $5 \cdot 32 = 160$ and a T-depth of $3 \cdot 32 = 96$.

4.1.1.2 Nielsen and Chuang Mapping

According to *Nielsen and Chuang* [98], if the ancilla line in the previous transformation is assigned to the 0 state, the fourth gate in (4.8) can be omitted. This approach leads to circuits with cheaper quantum cost, however, if there is no free line set to the 0 state, an additional line is required. The Nielsen and Chuang mapping results in an NCT circuit that consists of the following number of 2-control Toffoli gates

$$6(c - 3),\tag{4.12}$$

an NCV-cost of

$$18(c - 3) + 6, \tag{4.13}$$

and a T-depth of

$$18(c - 3). \tag{4.14}$$

In the remainder of this book, we denote the one ancilla mapping algorithms proposed by Nielsen and Chuang with NC.

Example 4.5. Figure 4.5b shows the resulting circuit after applying the NC transformation to the 7-control MCT gate depicted in Fig. 4.5a. Each of the resulting three MCT gates could be mapped using the B2 algorithm. The final result is depicted in Fig. 4.5c. Using an ancilla line set to 0 leads to a circuit with $6 \cdot (7 - 3) = 24$ Toffoli gates, an NCV-cost of $5 \cdot 24 = 120$, and a T-depth of $3 \cdot 24 = 72$. For the same MCT gate, it is obvious that using an arbitrary valued ancilla in the B1 mapping produces an NCT circuit with larger number of Toffoli gates and hence results in an expensive quantum cost.

4.1.1.3 Miller et al. Mapping

Miller et al. [91] have proven that an MPMCT gate can be mapped directly to a quantum circuit by making use of V gates. In particular, a Toffoli gate $T(C, t)$ may be mapped to the following circuit:

$$V(a_1, t) \circ T(C_1, a_1) \circ V^\dagger(a_1, t) \circ T(C_2, a_1) \circ V(a_1, t) \circ T(C_1, a_1)$$

$$\circ V^\dagger(a_1, t) \circ T(C_2, a_1) \tag{4.15}$$

In this case, the control set C_2 has one fewer line, at the expense of 4-control $V/V^\dagger$ gates. An ancilla is required, but similar to the Barenco et al. approach any free line may be used rather than a new one set to the 0 state. A c-control MPMCT gate can be realized, if $c \geq 5$, with at most

$$8(c - 3) \tag{4.16}$$

Toffoli gates. In addition, four NCV gates are required. Thus the obtained circuit, if the B2 is applied as a second step, has an NCV-cost of

$$24(c - 4) + 12 \tag{4.17}$$

and a T-depth of

$$24(c - 4). \tag{4.18}$$

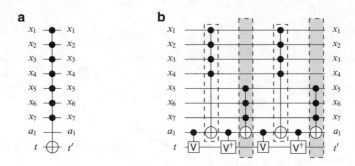

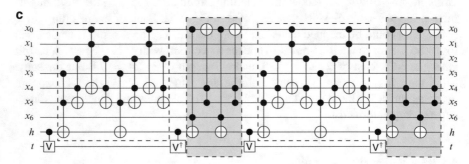

Fig. 4.6 Mapping an MPMCT gate using the Miller et al. mapping. (**a**) 7-control gate. (**b**) Miller et al. mapping. (**c**) Final mapping based on Miller et al. mapping

Table 4.1 Number of NCT gates and quantum cost of a c-control MPMCT gate

Mapping	Ancillae	NCT	NCV-cost	T-depth
Barenco et al. (Lemma 7.2) [15, 83]	$c - 2$	$4(c - 2) + 4$	$12(c - 2) + 2$	$12(c - 2)$
Barenco et al. (Lemma 7.3) [15]	1	$8(c - 3) + 4$	$24(c - 3) + 8$	$24(c - 3)$
Nielsen et al. [98]	1^a	$6(c - 3) + 4$	$18(c - 3) + 6$	$18(c - 3)$
Miller et al. [91]	1	$8(c - 4) + 8$	$24(c - 4) + 12$	$24(c - 4) + 8$

^{a}The ancilla is initialized to 0

In the remainder of this book, we denote the one ancilla mapping algorithms proposed by Miller et al. with MI.

Example 4.6. An example of the MI transformation is illustrated in Fig. 4.6b. The obtained circuit consists of $8(7 - 4) = 24$ Toffoli gates, and it has an NCV-cost of $40(7 - 4) + 4 = 124$ and a T-depth of $24(7 - 4) + 8 = 80$.

Table 4.1 summarizes the number of NCT gates, the NCV-cost, and the T-depth for the B1, NC, and, MI when they are applied to a c-control MPMCT gate with $c \geq 5$. In particular, for the one ancilla mappings, these metrics are calculated taking into account that the B2 is considered as a second step in the mapping of MPMCT gates. Note that the mapping of a c-control MPMCT gate with all negative controls and $(c - 2)$ ancillae will require only 2 additional NOT gates. Furthermore, the

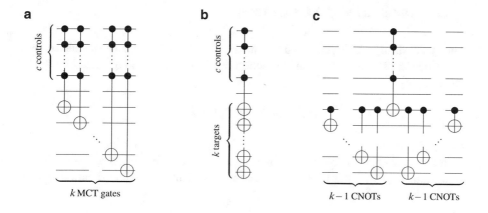

Fig. 4.7 Mapping of multiple targets gate. (**a**) MCT gates with common controls. (**b**) Multiple target gate. (**c**) New mapping

mapping of a c-control MPMCT gate with all negative controls will require 8, 6, and 8 additional NOT gates using the B1 algorithm, the NC algorithm, and the MI algorithm, respectively.

4.1.1.4 Miscellaneous

It has been shown in [80, 150] that the implementation of k c-control MCT gates with common controls can be done with an NCV-cost of

$$2(k-1) + 24(c-3) + 8 \qquad (4.19)$$

and a T-depth of

$$24(c-3). \qquad (4.20)$$

Note that the cost is derived from the B1 mapping algorithm. In fact, a circuit of MPMCT gates with the same controls (Fig. 4.7a) can be presented by only one gate known as *multiple target* gate (Fig. 4.7b). To implement this circuit, we need $(2k-2)$ CNOTs and one MPMCT gate with c-control. Therefore, the NCV-cost is $2k - 2 + 24(c-3) + 8 = 2k + 24(c-3)$ instead of $k(24(c-3) + 8)$.

Sasanain et al. [86, 110, 112] proposed techniques to reduce the NCV-cost of the MI mapping strategy by rearranging the variables of C_1 and C_2 and performing post-mapping optimizations.

4.1.2   Complexity of NCT Circuits

The synthesis algorithm presented by Shende et al. in [122] provides a constructive upper bound such that a reversible function over n variables can be realized with at most

$$c_{\text{NCT}_{B1}}(n) \leq 9n2^n + o(n2^n) \tag{4.21}$$

NCT gates.

A stronger upper bound can be found using the transformation based synthesis algorithm. In [76], it has been proven that an n-variable reversible function can be realized with at most

$$c_{\text{NCT}_{B1}}(n) \leq 5n2^n + o(n2^n) \tag{4.22}$$

NCT gates when the B1 mapping is applied.

4.2   Improving the Mapping of Single-Target Gates

Albeit providing a high-level representation for reversible circuits, the lower bound of the size of a reversible circuit consisting of Toffoli gates is exponential as we have seen in the last chapter. In other words, for every number of variables there exists a reversible function for which the size of the minimal circuit is exponential. In order to avoid this complexity when addressing large reversible functions and circuits, recently single-target gates are considered as a representation for reversible circuits because of their linear complexity as shown in [39]. Besides that, synthesis approaches presented in [39, 135] are based on this gate representation. However, for technology mapping into quantum circuits, so far single-target gates are mapped into circuits of Toffoli gates which are then independently mapped using the Barenco et al. mapping described in Sect. 4.1.1.

In the following, we describe a new mapping flow for determining quantum gate realizations for single-target gates. Since each single-target gate contains a Boolean control function, our method attempts to break large single-target gate into smaller ones using additional lines. As shown by our experimental evaluations, working on the higher level abstraction allows significant cost reductions. In the best case, we were able to reduce the costs of the quantum circuit by 75 % and in the average by about 20 % for both the NCV and the Clifford $+$ T gate libraries. The mapping method has been published in [5].

4.2.1 Motivation

So far, there is no mapping approach into quantum circuits that directly targets the single-target gates as it is done for the MCT or MPMCT gates. To map single-target gates, we aim at decomposing them into MPMCT gates so that we can afterwards map each obtained MPMCT gate using the approach explained above.

The mapping of a single-target gate to an MPMCT circuit is so far done by computing the ESOP expression of its controlling function, then each cube in the obtained expression is represented by an MPMCT gate as outlined in Fig. 4.8a.

Many Boolean decompositions, that have been summarized in Sect. 2.2, show their efficiency [94] in reducing the complexity of a Boolean function of n variables by decomposing it to simpler Boolean functions of less than n variables. Motivated by this, we aim at studying the impact of applying different kinds of Boolean decomposition while mapping single-target gates to MPMCT gates. Unlike the standard mapping, before determining the ESOP expression of the controlling function, we simplify such function using the Boolean decomposition, e.g., bi-decomposition, Ashenhurst, and Curtis decomposition, etc. Our new mapping flow for single-target gate based circuits is sketched in Fig. 4.8b. First, the ST gate is split to three simpler ST gates by decomposing its controlling function f to simpler sub-functions g_1, g_2, and g_3 using the Boolean decomposition. Then, each ST gate is transformed to an MPMCT subcircuit by determining the ESOP expression of its controlling function and substituting each cube with an MPMCT gate. Finally each MPMCT gate is mapped into an NCT circuit using the Barenco et al. algorithm reviewed in Sect. 4.1. Note that, in this example we used two ancilla set to 0 to compute the ST gates with the controlling functions g_1 and g_2, respectively. To restore the initial value of each ancilla, we reapply the two gates.

4.2.2 Mapping of Single-Target Gates

This section describes how Boolean decomposition can be applied to map reversible circuits composed of single-target gates into quantum circuits. Only the Young subgroup synthesis, for both the truth table based variant [39] and the BDD-based variant [135], makes use of single-target gates. However, due to the complexity of reversible circuits based on Toffoli gates (see, e.g., [4]) single-target gates are a preferable choice especially for large circuits.

Figure 4.9 shows how the functional decomposition of a single-target gate's control function can be used to generate less complex circuits. In the following we assume that the control function of the single-target gate that should be mapped depends on the variables $x_1, \ldots, x_{n-1}$.

Figure 4.9a shows the mapping approach for a disjoint Ashenhurst-Curtis decomposition. The variables are partitioned into four sets of variables represented as X_1, X_2, X_3, and X_4. First, the inner functions g_1, g_2, and g_3 are computed

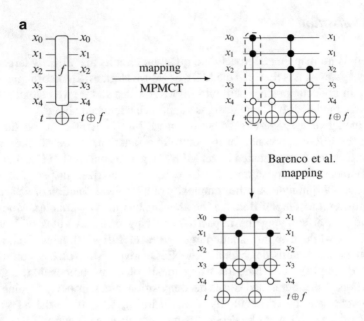

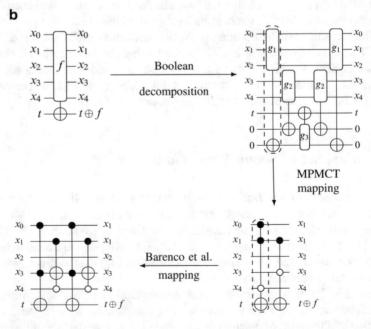

Fig. 4.8 Mapping flows for single-target gate based circuits. (**a**) Mapping flow for a single-target gate. (**b**) Proposed mapping flow for a single-target gate

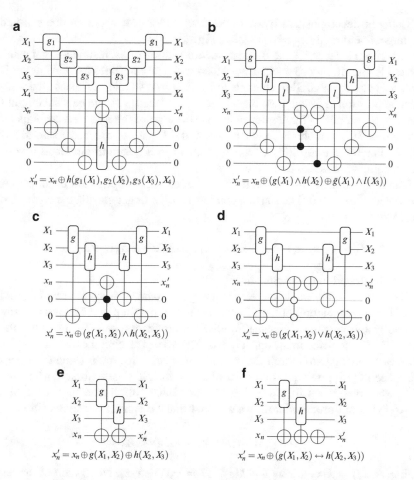

Fig. 4.9 Different types of decomposition. (**a**) Ashenhurst-Curtis. (**b**) MUX. (**c**) AND ($\wedge$). (**d**) OR ($\vee$). (**e**) XOR ($\oplus$). (**f**) XNOR ($\leftrightarrow$)

and each of their results is stored on an additional ancilla that is initialized
with a constant 0 value. Having the resulting values on these lines the outer
function can be computed and afterwards the constant values on the ancillae are
restored by reapplying the inner functions. Also, a decomposition based on the
MUX operation can be analogously performed by using 3 ancillae set to 0 (see
Fig. 4.9b). Figure 4.9c, d show non-disjoint bi-decomposition based on the AND
and OR operation, respectively. The sub-function f depends on variables in X_1 and
X_2 and the sub-function g depends on variables in X_2 and X_3. As can be seen,
the construction follows the representation of the Ashenhurst-Curtis decomposition
shown in Fig. 4.9a. Whether a decomposition is disjoint or non-disjoint does not
have an effect on the circuit construction but only on the size of the single-target
gates in terms of their support.

Using bi-decomposition based in the XOR and XNOR operator, one can update the target line directly as can be seen in Fig. 4.9e, f.

Algorithm 3 describes the Boolean decomposition of each single-target gate in a reversible circuit. First, 3 ancillae are added to the circuit (lines 3–6). Afterward, for each ST gate, the algorithm attempts to break it into smaller ST gates (lines 12). If a decomposition is found, then the ST gate is replaced with the new subcircuit that contains the smaller ST gates depending on the type of the retrieved decomposition (lines 15–37). After this step, each obtained ST gate is mapped to an MPMCT circuit. Finally, the resulting circuit is mapped to a quantum circuit using one of the mapping approaches defined in Sect. 4.1.

The remainder of this section discusses an example application of the approach illustrated in Fig. 4.10. The starting point is a single-target gate that is controlled by the control function

$$f(x_1, x_2, x_3, x_4) = \bar{x}_1 \bar{x}_2 \bar{x}_3 \bar{x}_4 \vee \bar{x}_1 x_2 \bar{x}_3 x_4 \vee x_1 \bar{x}_2 x_3 \bar{x}_4 \vee x_1 x_2 x_3 x_4$$

as also illustrated in its specification.

Decomposing the single-target gate using the standard mapping requires finding an ESOP representation of the function. One can obtain a smaller one in terms of literals by applying the *exorcism* [93] ESOP minimization tool. The resulting Toffoli gate circuit is depicted in the upper box of Fig. 4.10. The circuit consists of 4 Toffoli gates each having two controls. Mapping it into quantum circuits using the algorithm presented in [15] gives quantum costs of 21 for the NCV gate library and a T-depth of 12 when using the Clifford $+ T$ gate library. Each Toffoli gate has a T-depth of 3.

Applying our proposed flow, we will first find a disjoint bi-decomposition

$$f(x_1, x_2, x_3, x_4) = g(x_1, x_3) \wedge h(x_2, x_4)$$

with $g(x_1, x_3) = \bar{x}_1 \bar{x}_3 \oplus x_1 x_3$ and $h(x_2, x_4) = \bar{x}_2 \bar{x}_4 \oplus x_2 x_4$. Next, each of the resulting single-target gates controlled by g and h is mapped to Toffoli circuits as it is shown in the reversible circuit depicted in Fig. 4.10b. Each ST gate has two Toffoli gates with only one control while the last gate computes the AND of both subfunctions using a Toffoli gate with two controls. Finally, the resulting reversible circuit is mapped to a quantum circuit with the same algorithm used in the standard flow. The number of NCV gates of the resulting circuit is 13 (compared to 21) and the T-depth is 3 (compared to 12).

4.2.3 Experimental Evaluation

In order to confirm the benefits of incorporating the Boolean decomposition technique into the mapping flow of reversible circuits to quantum circuits described in Sect. 4.2.2, we have implemented the proposed idea in the open source toolkit *RevKit* [129]. The starting point is reversible circuits obtained from applying

---

Algorithm 3: Boolean decomposition of an ST circuit

Input: Reversible ST circuit G
Output: Reversible ST circuit G'

1 $k \leftarrow$ Size(G)
2 $i \leftarrow 0$
   // Adding 3 ancillas set to 0 to the new circuit G'
3 **while** $i \leq 3$ **do**
4   addAncilla(a_i, G')
5 $i \leftarrow i + 1$
6 **end**
7 $i \leftarrow 0$
8 **while** $i < k$ **do** // Terminate?
9 $g \leftarrow$ Gate(G, i)
10 $t \leftarrow$ target(g) // the target of the ST gate
11 $type \leftarrow -1$ // the type of the decomposition
      // find a decomposition for the controlling function of g.
      If found, G" will contain the small ST gates
12 $G'' \leftarrow$ BooleanDecompostion$(g, type)$
13 $g_1 \leftarrow$ Gate$(G'', 1)$
14 $g_2 \leftarrow$ Gate$(G'', 2)$
15   **if** $type =$ AND **then**
        // g_3 implements the AND operation
16 $g_3 \leftarrow T(\{a_1, a_2\}, t)$
17 $G' \leftarrow G' \circ g_1 \circ g_2 \circ g_3$
        // restore the value of the ancillas a_1 and a_2
18 $G' \leftarrow G' \circ g_1 \circ g_2$
19   **else if** $type =$ OR **then**
        // g_3 implements the OR operation
20 $g_3 \leftarrow T(\{a_1, a_2\}, t)$
21 $G' \leftarrow G' \circ g_1 \circ g_2 \circ g_3$
        // restore the value of the ancillas a_1 and a_2
22 $G' \leftarrow G' \circ g_1 \circ g_2$
23   **else if** $type =$ XOR **then**
24 $G' \leftarrow G' \circ g_1 \circ g_2$
25   **else if** $type =$ XNOR **then**
        // g_3 implements the NOT operation
26 $g_3 \leftarrow T(\emptyset, t)$
27 $G' \leftarrow G' \circ g_1 \circ g_2 \circ g_3$
28   **else if** $type =$ MUX **then**
29 $g_3 \leftarrow$ Gate$(G'', 3)$
        // g_4 implements the MUXER operation
30 $g_4 \leftarrow T(\{a_1, a_2\}, t) \oplus T(\{\bar{a}_1, a_3\}, t)$
31 $G' \leftarrow G' \circ g_1 \circ g_2 \circ g_3$
        // restore the value of the ancillas a_1 and a_2
32 $G' \leftarrow G' \circ g_1 \circ g_2$
33   **else** // the controlling function of the ST gate g could not
      be decomposed
34 $G' \leftarrow G' \circ g$
35   **end**
36 $i \leftarrow i + 1$
37 **end**

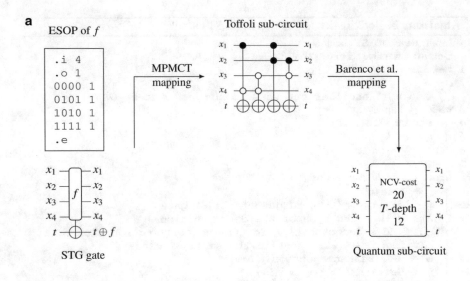

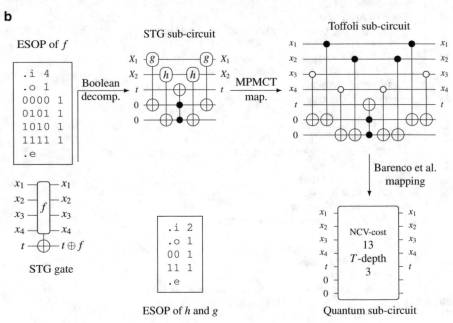

Fig. 4.10 Example of mapping a single-target gate. (**a**) Mapping example using the classical mapping flow. (**b**) Mapping example using the new mapping flow

the BDD-based version of the Young subgroup synthesis [135], which creates reversible circuits composed of single-target gates. We used the BDS-PGA tool [142] to decompose each control function of a single-target gate to smaller ones. We restricted the decomposition of each single-target gate to at most three smaller single-target gates in order to limit the use of additional lines to at most 3. To map the resulting smaller gates into circuits of Toffoli gates we used the ESOP minimization algorithm implemented in EXORCISM [93]. Finally, we applied the Barenco et al. mapping algorithm explained in [15].

4.2.3.1 Quantum Cost Evaluation

Experimental results are shown in Fig. 4.11a. In the graph, the NCV-cost obtained after applying the classical mapping is shown in white bar graph and the NCV-cost derived from the mapping algorithm based on Boolean decomposition is outlined in dotted bar graph. One can clearly see the efficiency of our new mapping approach on reducing the quantum cost. The same observation can be made for the T-depth depicted in Fig. 4.11b.

Detailed results are reported in Table 4.2. All benchmarks, names, and original lines are listed in the first and second columns, respectively. Then, the number of lines (L), the number of gates (GC), the NCV-cost (NCV), the T-depth (TD), and the required run-times in seconds (Time) are provided for the synthesized circuits based on standard mapping and the synthesized circuits based on Boolean decomposition are explained in Sect. 4.2.2.

We provide absolute and relative improvement in the last two columns for quantum costs in terms of the NCV and the Clifford + T gate libraries. The NCV-cost improvement of the circuits obtained by the proposed technique with respect to the realized circuits without taking into account the Boolean decomposition is given in the columns denoted by I_{NCV}. The procedure presented above yields circuits with lower NCV-cost comparing to circuits obtained by standard mapping. Table 4.2 shows an improvement percentage in terms of NCV-cost by approx. 20 % on average, 9 % in the worst case (*bw*), and in the best case (*cycle2*) by 48 %.

The T-depth cost reductions and its relative improvement are provided in the columns denoted by I_{TD}. Also for this gate library, realizations with fewer T-depth are achieved by applying our technique. The T-depth is reduced by 24 % on average, 9 % in the worst case (*bw*), and in the best case (*mod5d1*, and *mod5d2*) by 75 %.

4.2.3.2 Time Evaluation

Figure 4.12 sketches the run-time required by the classical mapping algorithm (white bar graph) and its equivalent required by the Boolean decomposition based mapping (dotted bar graph). The benchmarks are sorted in an increasing order with respect to the quantum cost (NCV-cost and T-depth) (see Table 4.2 to read the corresponding metrics for each benchmark). The run-time for both approaches is correlated with the quantum cost. Furthermore, although the Boolean decomposition based mapping run-time is higher than the one required by the classical mapping approach, the difference is not too large.

a

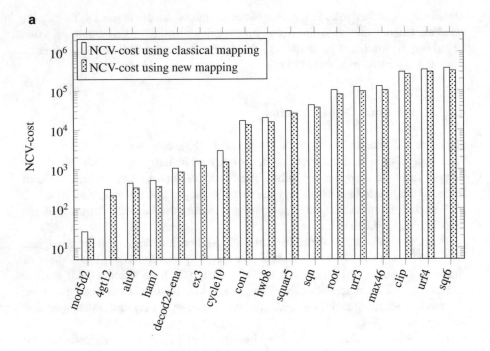

b

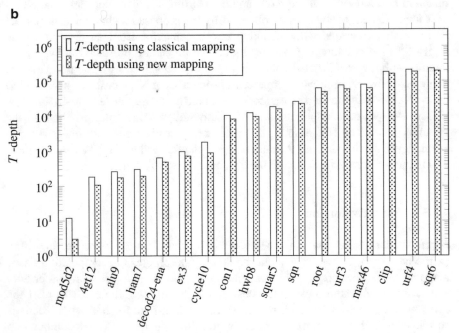

Fig. 4.11 Quantum cost resulting from classical and proposed approaches. (**a**) NCV-cost evaluation. (**b**) T-depth evaluation

Table 4.2 Experimental results for mapping ST circuits

Original benchmark (OB)		Classical mapping					Boolean decomposition based mapping						Improvements	
ID	L	L	GC	NCV	TD	Time	L	GC	NCV	TD	Time		I_{NCV} (%)	I_{TD} (%)
mod5d1	5	5	4	20	12	0.00	8	9	13	3	0.04		35.00	75.00
mod5d2	5	5	8	26	12	0.05	8	13	17	3	0.07		34.62	75.00
4gt12	5	5	16	310	183	0.06	8	40	214	108	0.19		30.97	40.98
alu9	5	5	21	449	261	0.09	8	72	337	174	0.30		24.94	33.33
4mod5	5	5	22	315	186	0.08	8	33	218	120	0.21		30.79	35.48
decod24-en.	6	6	28	1076	642	0.12	9	69	856	486	0.27		20.45	24.30
ex3	6	6	46	1614	963	0.16	9	96	1254	717	0.35		22.30	25.55
C17	6	6	51	1642	975	0.15	9	68	1344	789	0.37		18.15	19.08
ham7	7	7	44	525	300	0.12	10	85	368	192	0.32		29.90	36.00
con1	8	8	233	17,138	10,260	0.50	11	238	13,440	8043	1.07		21.58	21.61
z4	8	8	266	18,520	11,076	0.57	11	273	14,573	8721	1.04		21.31	21.26
hwb8	8	8	284	20,389	12,204	0.51	11	284	15,895	9516	0.93		22.04	22.03
wim	9	9	276	29,438	17,640	1.38	12	309	25,543	15,291	1.74		13.23	13.32
squar5	9	9	295	30,283	18,156	1.34	12	313	26,201	15,696	1.78		13.48	13.55
sqn	9	9	488	42,731	25,602	1.89	12	488	37,287	22,338	2.51		12.74	12.75
5xp1	10	10	806	103,052	61,806	5.33	13	818	81,111	48,639	6.31		21.29	21.30

(continued)

Table 4.2 (continued)

Original benchmark (OB)		Classical mapping					Boolean decomposition based mapping					Improvements	
ID	L	L	GC	NCV	TD	Time	L	GC	NCV	TD	Time	I_{NCV} (%)	I_{TD} (%)
root	10	10	847	103,721	62,196	4.09	13	847	80,907	48,516	5.45	22.00	21.99
max46	10	10	1060	130,699	78,366	4.14	13	1060	103,301	61,938	5.23	20.96	20.96
urf3	10	10	1081	124,847	74,865	2.97	13	1081	96,755	58,017	4.13	22.50	22.50
life	10	10	1137	135,935	81,492	4.39	13	1137	107,771	64,608	5.40	20.72	20.72
sym9	10	10	1175	134,692	80,742	4.37	13	1186	106,806	64,029	5.63	20.70	20.70
clip	11	11	2225	302,280	181,314	23.80	14	2225	266,754	160,002	25.52	11.75	11.75
urf4	11	11	2641	344,383	206,559	14.07	14	2641	302,907	181,683	18.22	12.04	12.04
cycle10	12	12	27	2972	1782	0.12	15	102	1536	879	0.29	48.32	50.67
sqr6	12	12	1750	375,060	225,012	45.68	15	1750	324,256	194,532	49.73	13.55	13.55
cm85a	13	13	9676	1,883,987	1,130,277	334.97	16	9676	1,698,685	1,019,109	378.69	9.84	9.84
add6	13	13	10,958	1,968,777	1,181,121	399.63	16	10,958	1,769,153	1,061,361	430.70	10.14	10.14
0,410,184	14	14	189	12,399	7428	0.20	17	197	10,695	6405	0.43	13.74	13.77
alu2	14	14	16,697	4,013,210	2,407,842	3273.48	17	16,697	3,400,136	2,040,006	3420.45	15.28	15.28
alu3	14	14	17,933	4,164,550	2,498,577	3422.97	17	17,939	3,501,451	2,100,744	3573.99	15.92	15.92
ham15	15	15	734	32,355	19,314	16.07	18	723	28,133	16,761	18.77	13.05	13.22
bw	32	32	2093	2,170,056	1,302,000	1202.30	35	2535	1,960,701	1,176,321	1399.81	9.65	9.65
Average												20.45	24.31

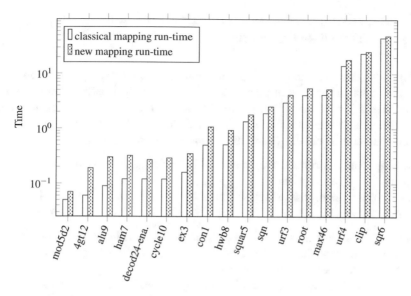

Fig. 4.12 Time evaluation

4.2.4 Remarks and Observations

Applying the BDS-PGA tool [142] to find a decomposition for an ST gate control function, it first searches for an algebraic decomposition and only looks for a Boolean decomposition if the first attempt is not successful. This process is performed recursively for each resulting sub-function until Boolean functions with at most two inputs are reached. We modified the tool such that recursion stops after maximum three decomposition in order to keep a reasonable number of additional lines.

We refer to the *common set* as the intersection of the bound set and free set in Ashenhurst-Curtis decomposition and as the intersection of supports in bi-decomposition. We have observed that for large common sets the results of the standard mapping approach outperform our approach. To ensure good results, we only decomposed functions with a small common set. We also noticed that bi-decomposition based on XOR and XNOR is less effective compared to the standard mapping since EXORCISM finds efficient ESOP representations. Consequently, we adapted the BDS-PGA tool such that it does not try to find bi-decomposition based on XOR or XNOR.

Finally, factorization with respect to a single variable cannot usually improve the overall result since the incorporation of the variable on the ancilla does not minimize the functional support. We have therefore turned off this option in the BDS-PGA tool.

To summarize, we looked for function decomposition with a small common set and allowed bi-decomposition only for the OR, AND, and MUX operator.

4.3   Improving the Mapping of MPMCT Gates
to Clifford + T Circuits

While automated Clifford + T optimization techniques exist [11], no optimized
mapping approaches with respect to the T-depth have yet been developed for the
Clifford + T library. Therefore, in this section, we present mapping schemes based
on existing algorithms to produce circuits of low T-depth. No previous mapping
scheme has considered this quantum cost metric explicitly. We present

- an improved algorithm to map c-control MPMCT gates into Clifford + T circuits
  using $(c - 2)$ ancillae, and
- an improved algorithm to map c-control MPMCT gates into Clifford + T circuits
  using one ancilla

These algorithms are then used to map reversible circuits into Clifford + T circuits,
taking into account the ancillae available at each reversible gate. The mapping
approaches have been published in [8].

This optimized reversible circuit mapping approach, integrated into our design
flow for quantum circuits, allows for significant T-depth reduction compared to
existing mapping approaches and Clifford + T circuit optimization algorithms. As
confirmed by an experimental evaluation, improvements of the T-depth of up to
65 % can be observed. This clearly demonstrates the efficiency of our approaches
on optimizing the cost of Clifford + T circuit.

4.3.1   Clifford + T Aware Reversible Circuit Mapping

Many algorithms have been proposed to map MPMCT gates as described in
Sect. 4.1.1 that mainly address the reduction of the quantum cost of quantum
circuits based on the NCV library [15, 75, 91, 111, 150]. Comparatively, no method
targeted the optimization of mapping approaches for the Clifford + T library. So
far Clifford + T circuits have only been considered in pre- and post-mapping
optimizations, e.g., [11], an optimization algorithm is given which takes mapped
NCT circuits and reduces their T count and depth. The work presented in [92]
reduces the T-depth of quantum circuits by applying optimizations either at the
reversible, pre-mapping, or quantum level. In [120], the author has given a class of
circuits with a T-depth of one by using a sufficient number of ancillae.

Following the previous observations, we propose mapping schemes that for the
first time take into account the Clifford + T gate library and the T-depth cost metric.

Algorithm 4: Mapping algorithm

Input: Reversible MPMCT based circuit G
Output: Quantum Clifford + T based circuit G'

```
1  k ← Size(G)
2  i ← 0
3  while i < k do
4      g ← Gate(G, i)
5      c ← Size(Controls(g))
6      a ← Size(Ancillas(g))
7      if c < 2 then
8          │  G' ← G' ∘ g
9      else if c = 2 then
10         │  G' ← G' ∘ ToffoliMapping(g)
11     else
12         │  if a ≥ ⌈(c+1)/2⌉ then
13         │      │  G' ← G' ∘ MultipleAncillasMapping(g)
14         │  else
15         │      │  G" ← ∅
16         │      │  G" ← OneAncillaMapping(g)
17         │      │  j ← 0
18         │      │  while j < 4 do
19         │      │      │  g ← Gate(G", j)
20         │      │      │  G' ← G' ∘ MultipleAncillasMapping(g)
21         │      │      │  j ← j + 1
22         │      │  end
23         │  end
24     end
25     i ← i + 1
26 end
```

4.3.2 Proposed Mapping Approaches

At a high level, our mapping scheme proceeds by first deciding on a mapping from a c-control MPMCT gate to Toffoli gates based on the number of available ancillae. Then, we use an optimized version of the mapping to get an efficient Clifford + T circuit. In particular, as sketched in Algorithm 4, for a given reversible circuit on n lines consisting of gates $g_i = \mathrm{T}(C_i, t_i)$ the algorithm maps each gate according to the following case distinction, using $c = |C_i|$ to refer to the number of controls and $a = n - c - 1$ to refer to the number of free lines.

Case 1: ($c < 2$) The gate g_i is already contained in the Clifford + T gate library and is directly added to the quantum circuit (see line 8).

Case 2: ($c = 2$) The gate g_i is mapped to its (T-depth) optimal quantum circuit according to Fig. 4.1c (see line 10).

Case 3: $(a \geq \lceil \frac{c+1}{2} \rceil)$ We apply a mapping scheme based on the B2 mapping that returns quantum circuits with a T-depth of $4(c-1)$ (see lines 12–13). This case is detailed in Sect. 4.3.3.1.

Case 4: (*otherwise*)   We map the gate with respect to a mapping scheme based on the B1, NC, or MI mapping (see lines 15–16). This case is detailed in Sect. 4.3.3.2.

If there exists a gate $T(C_i, t_i)$ such that $|C_i| = n - 1$, we add one additional line to the circuit before starting the mapping. Thus, it is ensured that there exists at least one ancilla line when applying the mapping in Case 4.

4.3.3 MPMCT Gates Mapping

Table 4.3 summarizes the T-depth for B1, B2, NC, and MI mapping algorithms defined in Sect. 4.1.1 when applied to an MPMCT gate with c control lines for $c \geq 5$. These bounds are obtained by directly mapping each two-control Toffoli to an (optimal) T-depth three circuit. Likewise, we map the 1-control V to an optimal circuit with a T-depth of 2. In the following we derive better upper bounds by making use of T gate cancellations.

The remainder of this section describes our optimized mappings for each type of MPMCT circuit into Clifford $+ T$ gates.

4.3.3.1 MPMCT Gates Mapping Using $c - 2$ Ancillae

In the following we present a mapping scheme based on the B2 mapping, using $c-2$ ancillae. We present the mapping by applying series of rewrites to the B2 mapping.

Lemma 4.1. *A c-control MPMCT gate with $c \geq 4$ can be realized with a T-depth of*

$$8(c - 2). \tag{4.23}$$

Proof. We first note that the T-depth for a c-control MPMCT gate is the same as the T-depth for a c-control MCT gate. In particular, we may write the MPMCT gate

Table 4.3 T-depth for a c-control MPMCT gate

Mapping	Ancillae	Clifford $+ T$
B1 (Barenco et al. (Lemma 7.3 [15]))	1	$24(c - 3)$
NC (Nielsen and Chuang [98])	1[a]	$18(c - 3)$
MI (Miller et al. [91])	1	$24(c - 4) + 8$
B2 (Barenco et al. (Lemma 7.2 [15]))	$c - 2$	$12(c - 2)$

[a]The ancilla is initialized to 0

as an MCT gate by adding NOT gates. Since the NOT gates are irrelevant to the T-depth, we may thus calculate the T-depth for an MPMCT gate by calculating the T-depth of an MCT gate with the same number of controls.

We now show that the B2 mapping gives a T-depth of $8(c-2)$. To do so, we introduce the (2-control) ωX gate

$$\omega X : |xyz\rangle \mapsto e^{\frac{i\pi}{4}x}|xy(x \wedge y \oplus z)\rangle$$

where $\omega X^\dagger$ is inverse of the ωX gate. We may write a two-control Toffoli gate as the 2-control ωX with an added $T^\dagger$ gate:

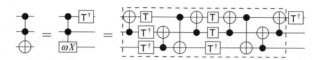

or equivalently as a $\omega X^\dagger$ gate with an additional T gate. Note that the circuit symbol for the ωX gate is ambiguous as to which control has the added phase rotation. For our purposes, it suffices to assume the top-most control always has the additional phase. Similar to the iX gates introduced in [120], the ωX gate serves as a convenient way to factor a $T/T^\dagger$ gate out of the two-control Toffoli, and observe the cancellation of T gates between adjacent controls. An algebraic treatment of this cancellation property is also possible [11].

We use T-gate cancellations to reduce every Toffoli in the B2 mapping of a multiple control Toffoli to a ωX or $\omega X^\dagger$ gate. As shown in Fig. 4.13, two control Toffolis with the same target are alternately mapped to either the ωX or the $\omega X^\dagger$ gate. Since no gate appears between each matching pair's top-most control, the $T^\dagger$ and T gates may then be cancelled. Thus, we see that the T-depth is given by $4(c-2)u_{\omega X}$, where $u_{\omega X}$ is the T-depth of the ωX gate, since the $\omega X^\dagger$ gate may be implemented by reversing and inverting the ωX gate. The ωX gate can be implemented in T-depth 2, as shown in Fig. 4.13d. Thus, a c-control MPMCT gate can be realized with T-depth $8(c-2)$ using the B2 mapping.   ∎

The above theorem assumes that no constant ancilla is added, except for those needed in the multiple control Toffoli decompositions. Recent results [11, 120] show that better T-depths can be achieved with a relatively small number of constant ancillae.

The previous bounds on T-depth looked at the T-depth required to implement a particular two-control Toffoli gate. However, it was shown in [3] that much more T-depth reduction is available if the number of Hadamard gates can be reduced. In this section, we consider these optimizations and give better MPMCT mapping technique. Note that in practice circuits of better T-depth can be found by moving phase gates between the Toffoli gate subcircuits [11].

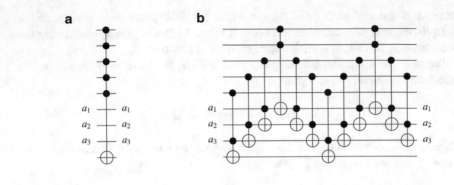

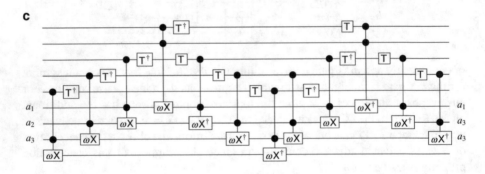

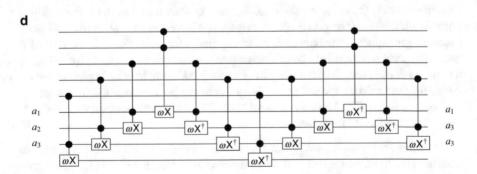

Fig. 4.13 Optimized B2 mapping with respect to T-depth. (**a**) 5-control gate. (**b**) B2 mapping (Lemma 7.2). (**c**) Toffoli gates replacement. (**d**) Optimized B2 mapping

Lemma 4.2. *A c-control MPMCT gate with $c \geq 4$ can be realized with a T-depth of*

$$4(c - 1) \tag{4.24}$$

using the B2 mapping with $c - 2$ ancillae.

Proof. We first note that the T-depth for a c-control MPMCT gate is the same as the T-depth for a c-control MCT gate. In particular, we may write the MPMCT gate as an MCT gate, conjugated on some bits by NOT gates. Since the NOT gates are irrelevant to the T-depth, we may thus calculate the T-depth for an MPMCT gate by calculating the T-depth of an MCT gate with the same number of controls.

Consider the c-control MPMCT gate mapped using the B2 algorithm as shown in Fig. 4.13b. Note that every Toffoli shares either two controls, or a control and a target with another Toffoli, with nothing in between. In particular, we may rewrite every Toffoli with a doubly controlled Z gate and two Hadamard gates on either side of the target line. For matched pairs of Toffolis sharing a target, the Hadamard gates cancel:

By rewriting every Toffoli with a 2-control Z gate and two Hadamard gates on either side of the target bit, we can remove each two Hadamard gates having same target (see Fig. 4.14a). Recall that controlled phase gates are symmetric in that the target behaves like a control [133], i.e.,

By applying this fact (see Fig. 4.14b), we may observe that each 2-control Z gate now shares exactly two controls with another gate, a fact that can be used to eliminate T gates by using the iZ-gate [120], defined as

$$iZ : |xyz\rangle \mapsto \omega^{4xyz-2xy}|xyz\rangle$$

where $\omega = e^{\frac{i\pi}{4}}$. We denote the inverse of the iZ gate by $iZ^\dagger$. The iZ gate implements the 2-control Z gate up to some smaller phase using only 4 T-gates:

For each pair of 2-control Z gates that share two controls, one is written as an iZ gate and a 1-control $S^\dagger$ gate, and the other is written as an $iZ^\dagger$ and a 1-control S gate (see Fig. 4.14c)—the 1-control $S/S^\dagger$ gates then cancel (see Fig. 4.14d).

a

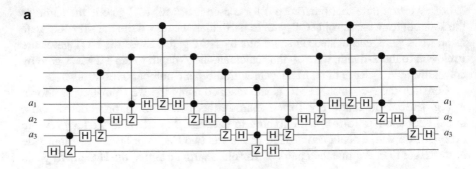

b

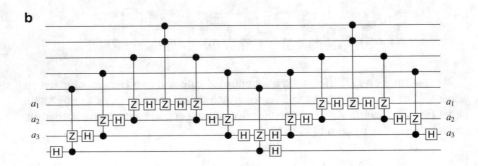

c

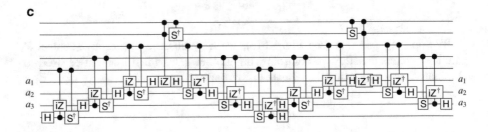

d

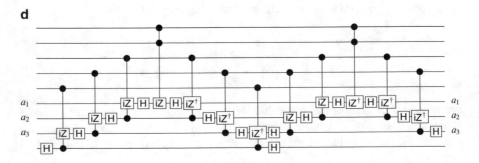

Fig. 4.14 Optimization of B2 mapping with respect to T-depth. (**a**) Toffoli gates replacement. (**b**) Swoping the target and a control of the Z gates. (**c**) Z gates replacement. (**d**) Redudant gates reduction

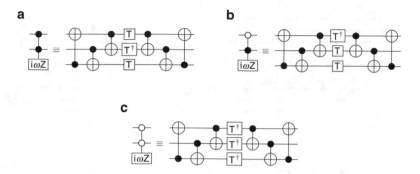

Fig. 4.15 Clifford+*T* implementations for mixed-polarity $i\omega Z$ gates. (**a**) iZ gate with two positive controls. (**b**) iZ gate with a negative and a positive control. (**c**) iZ gate with negative controls

To further optimize this mapping, we define a refinement of the iZ gate—the $i\omega Z$ gate

$$i\omega Z : |xyz\rangle \mapsto \omega^{4xyz-2xy-z}|xyz\rangle$$

where $\omega = e^{\frac{i\pi}{4}}$ and $i\omega Z^\dagger$ is its inverse. This gate may be implemented in T-depth 1 (see Fig. 4.15), and together with a $T^\dagger$ gate on the target implements the iZ gate:

Figure 4.16a shows the B2 mapping of the 5-control MCT gate where iZ gates have been replaced with $i\omega Z/i\omega Z^\dagger$ gates, and the T gates are either cancelled or parallelized (see Fig. 4.16b). Note that the 4 "points" of the circuit are left as iZ gates since they cannot be parallelized further. The extra phase gates from the outermost circuits (see Fig. 4.16b) are also cancelled despite being physically separated in the circuit, since each ancilla is returned to its initial state as shown in Fig. 4.16c.

To complete the analysis, we note that each $iZ/iZ^\dagger$ gate can be mapped to 4 T gates in depth 2 [120], while each $i\omega Z/i\omega Z^\dagger$ gate can be mapped into a quantum circuit with a T-depth of 1 (see Fig. 4.15). This gives a total T-depth of

$$4(c-2) + 4 = 4(c-1).$$

■

Note that this bound is tighter than the T-depths achieved by T-par [11], showing that their heuristic approach is nonoptimal in some cases. Moreover, using a single 0-valued ancilla, the T-depth may be reduced to $4(c-2)$. Specifically, using a single ancilla in the zero state, the $iZ/iZ^\dagger$ gates may be mapped to a circuit with a T-depth of 1 [120], giving a total T-depth of 1 for each two-control Toffoli in the original mapping—i.e., $4(c-2)$.

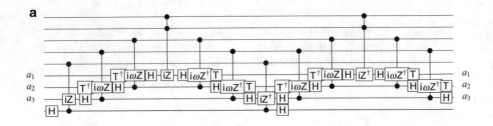

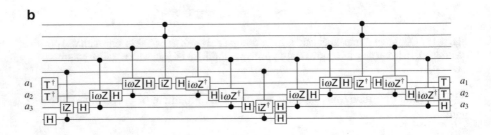

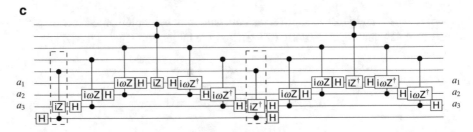

Fig. 4.16 T-depth of the 5-control MCT gate using B2 mapping. (**a**) iZ gate replacement. (**b**) Redundant gates reduction. (**c**) Final mapping result

Next, we give the T-depth of MPMCT gate using a single ancilla mapping algorithms B1, NC, and MI. For each, a c-control MPMCT gate is broken down into m-control and $c - m + 1$ or $c - m$-control MPMCT gates, which are further decomposed using the B2 mapping. In addition to the optimized B2 mapping explained above, we further reduce the T-depth by using the *self-inverse* property of MPMCT gates to cancel additional $T/T^\dagger$ gates.

4.3.3.2 MPMCT Gates Mapping Using One Ancilla

In this section we give optimized mapping schemes using one ancilla based on the B1, NC, and MI mappings. In each mapping algorithm, a c-control MPMCT gate is decomposed into m-control and $(c - m + 1)$-control (or $(c - m)$-control in the case of the MI mapping) MPMCT gates, which are further decomposed using the optimized B2 mapping explained above.

Lemma 4.3. *A c-control MPMCT gate with $c \geq 5$ controls can be realized with a T-depth of*

$$8(c - 2) \qquad \text{based on the B1 mapping,} \qquad (4.25)$$

$$6(c - 2) + 2 \qquad \text{based on the NC mapping, and} \qquad (4.26)$$

$$8(c - 3) + 4 \qquad \text{based on the MI mapping} \qquad (4.27)$$

Proof. Consider the B1 mapping, as in Fig. 4.4b. If we map the first two MPMCT gates using the B2 mapping above, then use the *inverse* (obtained by reversing the circuit and replacing each T gate with $T^\dagger$) of the B2 mapping for the remaining MPMCT gates, we can cancel an additional 8 T gates. Specifically, for each pair of MPMCT gates (first and third or second and fourth), the value on the target line of either dashed iZ gate shown in Fig. 4.16c is constant. As in the proof of the B2 upper bounds, we may then replace each dashed iZ gate with an $i\omega Z$ gate and cancel the extra T gates. The result is a total reduction of 8 levels of T-depth, giving a total T-depth for a c-control MPMCT gate of

$$2 \cdot 4(m - 1) + 2 \cdot 4(c + 1 - m - 1) - 8 = 8(c - 2).$$

Using the same argument for the NC mapping, additional $T/T^\dagger$ gate (and layer of T-depth) from the first and third MPMCT gates are removed. The resulting mapping has a T-depth of $6(c - 2)$ when c is even and $6(c - 2) + 2$ when c is odd.

Finally, consider the MI mapping. We use the same argument to reduce each of the four multiple control Toffolis by one layer of T-depth. We further note that each 1-control V gate may be mapped in T-depth 1. In particular, we first map each V to 1-control S gates conjugated by Hadamards

then cancel adjacent Hadamards. The 1-control S gates are then mapped alternately into $\omega S/\omega S^\dagger$ gates

where ωS maps $|xy\rangle$ to $\omega^{2xy-y}|xy\rangle$. As shown in the following:

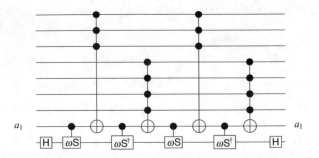

the additional $T/T^\dagger$ gates cancel, leaving 4 $\omega S/\omega S^\dagger$ gates, each of which is mapped to two $T/T^\dagger$ gates in T-depth 1. The total T-depth for the MI mapping scheme is then at most $2 \cdot 4(m-1) + 2 \cdot 4(c-m-1) - 8 + 4 = 8(c-3) + 4$. ∎

While the NC mapping clearly gives the best T-depth, it should be noted that it does so by using a zero-valued ancilla and so may require an extra line to be added where the B1 and MI mappings do not.

4.3.4 Experimental Results

In this section we present the experimental results of the proposed approaches for mapping mixed-polarity Toffoli gates into the Clifford $+ T$ gate library.

The mapping approaches above have been implemented in the open source toolkit *RevKit* [129] and tested on a suite of benchmarks taken from [73, 147]. The Benchmarks were embedded using the algorithm proposed in [136], then the resulting reversible functions were synthesized with the approach outlined in [135].

4.3.4.1 Mapping of MPMCT Gates

The experimental results presented graphically in the plot in Fig. 4.17a show the T-depth of mapped MPMCT gates with up to 15 controls based on B1. The values of x-axis and the y-axis denote the number of controls and the T-depth, respectively. The plot contains three different scenarios: the T-depth of the MPMCT gates mapped using the original B1 (blue line), the T-depth of the mapped MPMCT using B1 and then optimized with Tpar [11] (red line), and the T-depth of the MPMCT gates mapped using the optimized B1 (green line) as described in Sect. 4.3.2. The results show that our approach has achieved the smallest T-depth for each MPMCT gates in comparison with the two other mentioned methods. The overall improvement of the T-depth reaches 60 and 14 % with respect to the original B1 and Tpar, respectively. The same observation can be concluded for the B2 mapping as outlined in Fig. 4.17b.

Fig. 4.17 T-depth of mapped
MPMCT gates and circuits.
(**a**) T-depth of MPMCT gates
using B1. (**b**) T-depth of
MPMCT gates using B2
mapping. (**c**) T-depth of
MPMCT gates using one
ancilla mappings

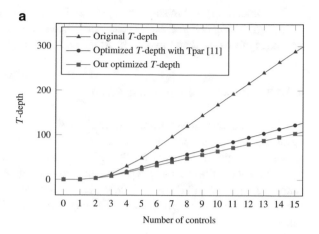

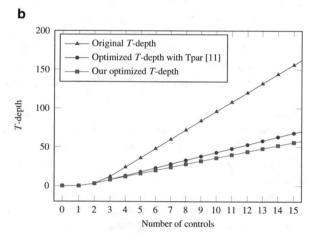

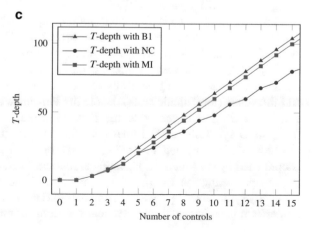

The plots in Fig. 4.17c show the optimized T-depth of mapped MPMCT for up to 15 controls using one ancilla. It is clear that the results of the optimized NC mapping approach are better than the optimized B1 and MI mapping approaches.

4.3.4.2 Mapping of Reversible Circuits

The obtained results are shown in Table 4.4. For each benchmark we show the name (ID), the number of lines (L), and the number of gates (GC). Then we apply the mapping algorithms B2, NC, and MI, respectively, when the gate has no sufficient ancillae to apply B1. For each case, we give the T-depth of the original mapping algorithm (TD_0), the T-depth of the improved mapping algorithm (TD_1), the required run-time in seconds (Time), and the T-depth improvement I_{TD}, respectively.

Experiments demonstrate the efficiency of the proposed mapping methodologies with respect to original mapping algorithms. Compared to the best previously introduced methods, we show that our mapping yields substantially smaller circuits. More precisely, improvements of around 57, 54, and 54 % can be achieved on average for B2, NC, and MI, respectively. In the best case, the T-depth of the circuits can even be reduced by more than 54 % for each mapping approach.

Next, we compared the efficiency of each mapping algorithm with respect to the obtained T-depth. The plot in Fig. 4.19 shows the T-depth of each circuit using the three different mapping strategies B2 (dashed-lines bar), NC (white bar), and MI (dotted bar). As expected, since the MI approach is an improvement of the B1 algorithm, we clearly see that MI outperforms B1. As shown in the experimental results, the NC algorithm beats each of B1 and MI algorithms, this is explained by the fact of omitting an MPMCT gate on the mapping due to the use of an ancilla line set to the constant value 0. An interesting question for future research is when the cost of adding extra 0-valued ancillae outweighs the extra T-depth incurred by using the B2 or MI approach.

Finally, we compare the following synthesis approaches in terms of the resulting T-depth: the QMDD based synthesis approach [131], the transformation based synthesis approach (TBS [88]), the Young subgroups based synthesis approach (YSG [135]), and the Reed-Muller synthesis approach (RMS [82]). The experimental results are shown graphically in the plots of Fig. 4.18. The values of x-axis and the y-axis (logarithmic scale) denote the benchmark and the T-depth. The plot contains four different scenarios: the T-depth of the mapped circuits generated by the QMDD synthesis (dashed-lines bar), the TBS synthesis (white bar), the YSG synthesis (dotted bar), and the RMS synthesis (black bar). Note that each benchmark is optimized by the simulated annealing approach. Thereafter, the optimized circuits are mapped using the NC algorithm. In most of the cases, the mapped circuits originally synthesized by the RMS approach [82] as well as the YSG approach [135] outperform the other synthesis techniques in terms of producing lower T-depth.

Table 4.4 Experimental results for mapping MPMCT circuits

Original benchmark (OB)			B1 [15] mapping				NC [98] mapping				MI [91] mapping			
ID	L	GC	TD_0	TD_1	Time	I_{TD} (%)	TD_0	TD_1	Time	I_{TD} (%)	TD_0	TD_1	Time	I_{TD} (%)
4_49_7	4	14	84	60	0.00	28.57	66	54	0.00	18.18	84	60	0.00	28.57
4gt10	5	19	333	185	0.00	44.44	207	143	0.00	30.92	233	145	0.00	37.77
decod24-enable	6	30	666	308	0.00	53.75	441	285	0.00	35.37	472	292	0.00	38.14
majority	6	70	1584	766	0.01	51.64	1071	703	0.00	34.36	1154	726	0.00	37.09
f2	7	96	3771	1831	0.01	51.45	2907	1505	0.01	48.23	2811	1591	0.00	43.40
sym6	7	163	5838	2878	0.02	50.70	4686	2444	0.01	47.84	4590	2566	0.01	44.10
z4	8	329	15,036	6260	0.02	58.37	11,682	5432	0.02	53.50	11,756	5868	0.02	50.09
hwb8	8	372	16,488	6976	0.03	57.69	12,954	6026	0.02	53.48	13,000	6520	0.02	49.85
wim	9	364	24,423	10,487	0.06	57.06	19,155	8445	0.03	55.91	20,583	9527	0.04	53.71
squar5	9	394	25,449	10,949	0.06	56.98	20,088	8912	0.03	55.64	21,593	9985	0.04	53.76
adr4	9	606	35,334	15,434	0.07	56.32	28,428	12,712	0.04	55.28	30,182	14,146	0.04	53.13
sqrt8	9	666	40,239	17,543	0.09	56.40	32,049	14,345	0.04	55.24	34,175	16,027	0.05	53.10
hwb9	9	807	41,673	18,481	0.08	55.65	34,182	15,512	0.04	54.62	36,105	17,089	0.05	52.67
dc1	10	273	28,596	11,352	0.07	60.30	21,474	8872	0.03	58.68	24,596	10,416	0.04	57.65
root	10	1135	88,671	33,847	0.16	61.83	68,529	28,579	0.11	58.30	75,087	31,639	0.13	57.86
dist	10	1283	101,067	38,383	0.17	62.02	78,081	32,475	0.10	58.41	85,515	35,915	0.14	58.00
sym9	10	1610	112,776	42,636	0.15	62.19	88,350	36,910	0.13	58.22	95,896	40,272	0.15	58.00
rd84	11	2901	241,866	98,882	0.36	59.12	193,320	81,350	0.31	57.92	214,922	92,146	0.37	57.13

(continued)

Table 4.4 (continued)

Original benchmark (OB)			B1 [15] mapping				NC [98] mapping				MI [91] mapping			
ID	L	GC	TD_0	TD_1	Time	I_{TD} (%)	TD_0	TD_1	Time	I_{TD} (%)	TD_0	TD_1	Time	I_{TD} (%)
clip	11	3138	252,207	103,551	0.41	58.94	202,599	85,727	0.33	57.69	224,223	96,555	0.41	56.94
sym10	11	3539	296,790	121,358	0.46	59.11	236,955	99,957	0.29	57.82	263,606	113,062	0.44	57.11
cm152a	11	3804	300,108	123,284	0.47	58.92	242,481	102,543	0.31	57.71	267,836	115,216	0.33	56.98
urf4	11	3831	305,040	125,236	0.47	58.94	246,180	104,070	0.31	57.73	272,048	116,988	0.34	57.00
plus63 4096	12	18	918	362	0.00	60.57	732	304	0.00	58.47	838	346	0.00	58.71
cycle10	12	27	1782	734	0.01	58.81	1419	579	0.00	59.20	1606	690	0.00	57.04
sqr6	12	2365	323,625	122,313	0.49	62.21	243,651	95,419	0.37	60.84	287,561	114,281	0.45	60.26
plus127 8192	13	19	1098	442	0.00	59.74	882	364	0.00	58.73	1018	422	0.00	58.55
plus63 8192	13	20	1146	462	0.01	59.69	930	384	0.00	58.71	1066	442	0.01	58.54
cm42a	13	73	13,056	4904	0.02	62.44	9792	3744	0.02	61.76	11,952	4628	0.02	61.28
dc2	13	5696	893,712	342,052	1.33	61.73	674,424	263,440	1.09	60.94	807,824	320,580	1.33	60.32
0,410,184	14	193	7707	3295	0.01	57.25	7683	3287	0.01	57.22	7707	3295	0.01	57.25
misex1	14	5242	1,013,910	377,946	1.43	62.72	760,881	289,451	1.17	61.96	931,174	357,574	1.43	61.60
urf6	15	2350	577,728	211,376	0.82	63.41	433,296	160,540	0.52	62.95	540,128	201,976	0.62	62.61
C7552	20	274	84,264	29,992	0.14	64.41	63,198	22,784	0.09	63.95	80,456	29,040	0.12	63.91
bw	32	3709	2,264,256	783,976	2.92	65.38	1,698,192	591,940	2.55	65.14	2,205,808	769,364	3.02	65.12
Average						57.92				54.38				54.75

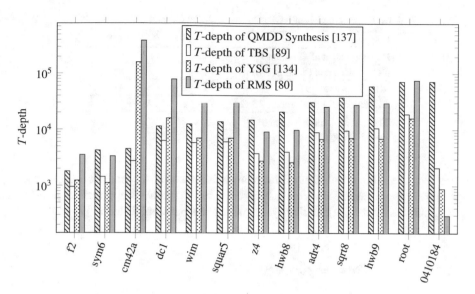

Fig. 4.18 *T*-depth of circuits obtained from different synthesis approaches

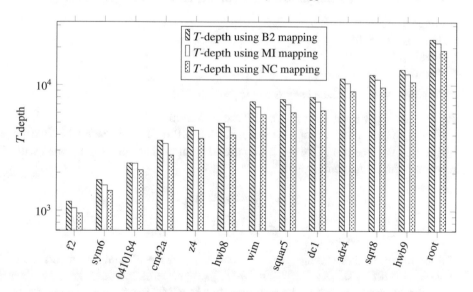

Fig. 4.19 *T*-depth of mapped circuit using one ancilla mapping algorithms

4.4 Complexity Analysis of NCT Circuits

In this section, we present tighter upper bounds on the number of NCT gates needed to realize an MPMCT gate, an ST gate, and a reversible circuit. Both multiple controlled Toffoli gates and mixed-polarity Toffoli gates have been considered for

Table 4.5 Number of NCT gates for an MPMCT gate with c controls

Mapping	Ancillae	NCT
B1 (Barenco et al. (Lemma 7.3 [15]))	1	$8(c-3)$
NC (Nielsen and Chuang [98])	1^a	$6(c-3)$
MI (Miller et al. [91])	1	$8(c-4) + 4^b$
B2 (Barenco et al. (Lemma 7.2 [15])	$c-2$	$4(c-2)$

[a]The ancilla is initialized to 0
[b]Four NCV gates

this purpose. The calculation of the bounds is based on the transformation based synthesis approach and the Young subgroups based synthesis approach that results in circuits using the ST gate library. The study has been published in [7, 134].

4.4.1 Upper Bounds for MPMCT Gates

Theorem 4.1. *A c-control MPMCT gate can be realized, if $c \geq 5$, with at most*

- $um_{\mathrm{NCT_{B1}}}(c) \leq 8(c-3)$
- $um_{\mathrm{NCT_{NC}}}(c) \leq 6(c-3)$
- $um_{\mathrm{NCT_{MI}}}(c) \leq 8(c-4) + 4$
- $um_{\mathrm{NCT_{B2}}}(c) \leq 4(c-2)$

NCT gates. Note that these bounds are upper bounds.

Proof. The upper bounds for B1 and B2 are proven in [15].

In the NC mapping two of the three resulting MPMCT gates have $\lceil \frac{c}{2} \rceil$ controls while the third one has $c + 1 - \lceil \frac{c}{2} \rceil$ controls. Each of these gates is mapped with respect to the B2 mapping, giving

$$um_{\mathrm{NCT_{NC}}}(c) \leq 2 \cdot 4 \left(\left\lceil \tfrac{c}{2} \right\rceil - 2 \right) + 4 \left(c + 1 - \left\lceil \tfrac{c}{2} \right\rceil - 2 \right)$$
$$\leq 4 \left(c + \left\lceil \tfrac{c}{2} \right\rceil - 5 \right).$$

For an even c, we have $um_{\mathrm{NCT_{NC}}}(c) \leq 6 \left(c - \frac{10}{3} \right) \leq 6(c-3)$, and for an odd c, we have $u(c) \leq 6(c-3)$.

In the MI mapping two of the resulting four MPCMCT gates have $\lceil \frac{c}{2} \rceil$ controls while the other two have $c - \lceil \frac{c}{2} \rceil$ controls. Applying B2 mapping yields

$$um_{\mathrm{NCT_{MI}}}(c) \leq 2 \cdot 4 \left(\left\lceil \tfrac{c}{2} \right\rceil - 2 \right) + 2 \cdot 4 \left(c - \left\lceil \tfrac{c}{2} \right\rceil - 2 \right)$$
$$\leq 8(c-4)$$

Toffoli gates. In addition, four NCV gates are required.   ∎

Table 4.5 summarizes the above upper bounds for the number of Toffoli gates given by each mapping from the previous section when applied to an MPMCT gate with $c \geq 5$ control lines. Note that for the MI mapping four additional NCV gates are included and that four additional NOT gates need to be added in case the MPMCT gate has only negative control lines.

4.4.2 Upper Bounds for Single-Target Gates

4.4.2.1 Upper Bounds Based on PPRM Expressions

Tighter upper bounds for reversible circuits can be obtained by combining the synthesis approach outlined in the previous section using MCT gates and upper bounds for the size of PPRM expressions. We make use of the following lemma:

Lemma 4.4.

$$\sum_{1 \leq i \leq \lceil \frac{n}{2} \rceil} i \binom{n}{i} = \begin{cases} n \cdot 2^{n-2} & n \text{ even} \\ n \cdot 2^{n-2} + \frac{n+1}{4} \binom{n}{\frac{n+1}{2}} & n \text{ odd} \end{cases}$$

$$\sum_{\lceil \frac{n}{2} \rceil < i \leq n} i \binom{n}{i} = \begin{cases} n \cdot 2^{n-2} & n \text{ even} \\ n \cdot 2^{n-2} - \frac{n+1}{4} \binom{n}{\frac{n+1}{2}} & n \text{ odd.} \end{cases}$$

Proof. Taking into consideration that

$$\sum_{i=1}^{n} i \binom{n}{i} = n \cdot 2^{n-1} \tag{4.28}$$

$$i \binom{n}{i} = (n - (i - 1)) \binom{n}{n - (i - 1)} \tag{4.29}$$

we consider the cases when n is even and n is odd. n **is even:** we have

$$\lceil \tfrac{n}{2} \rceil = \frac{n}{2}$$

Then

$$\sum_{i=1}^{\frac{n}{2}} i \binom{n}{i} + \sum_{i=\frac{n}{2}+1}^{n} i \binom{n}{i} = n \cdot 2^{n-1}$$

$$\sum_{i=1}^{\frac{n}{2}} (n-(i-1)) \binom{n}{n-(i-1)} + \sum_{i=\frac{n}{2}+1}^{n} i \binom{n}{i} = n \cdot 2^{n-1}$$

$$\sum_{i=\frac{n}{2}+1}^{n} i \binom{n}{i} + \sum_{i=\frac{n}{2}+1}^{n} i \binom{n}{i} = n \cdot 2^{n-1}$$

$$2 \sum_{i=\frac{n}{2}+1}^{n} i \binom{n}{i} = n \cdot 2^{n-1}$$

Therefore

$$\sum_{i=\frac{n}{2}+1}^{n} i \binom{n}{i} = n \cdot 2^{n-2}$$

$$\sum_{i=1}^{\frac{n}{2}} i \binom{n}{i} = n \cdot 2^{n-2}$$

n is odd: we have

$$\lceil \tfrac{n}{2} \rceil = \frac{n+1}{2}$$

Then

$$\sum_{i=1}^{\frac{n+1}{2}} i \binom{n}{i} + \sum_{i=\frac{n+3}{2}}^{n} i \binom{n}{i} = n \cdot 2^{n-1}$$

$$\sum_{i=1}^{\frac{n+1}{2}} (n-(i-1)) \binom{n}{n-(i-1)} + \sum_{i=\frac{n+3}{2}}^{n} i \binom{n}{i} = n \cdot 2^{n-1}$$

$$\frac{n+1}{2} \binom{\frac{n+1}{2}}{n} + 2 \sum_{i=\frac{n+3}{2}}^{n} i \binom{n}{i} = n \cdot 2^{n-1}$$

Hence

$$\sum_{i=\frac{n+3}{2}}^{n} i \binom{n}{i} = n \cdot 2^{n-2} - \frac{n+1}{4} \binom{\frac{n+1}{2}}{n}$$

$$\sum_{i=1}^{\frac{n+1}{2}} i \binom{n}{i} = n \cdot 2^{n-2} + \frac{n+1}{4} \binom{\frac{n+1}{2}}{n}$$

∎

Corollary 4.1. *From Lemma 4.4, we derive the following inequations for $n \geq 6$:*

$$4 \cdot \sum_{i=0}^{\lceil \frac{n}{2} \rceil} (i-2) \binom{n}{i} + 8 \cdot \sum_{i=\lceil \frac{n}{2} \rceil+1}^{n} (i-3) \binom{n}{i} \leq (6n-24) \cdot 2^{n-1} \qquad (4.30)$$

$$4 \cdot \sum_{i=0}^{\lceil \frac{n}{2} \rceil} (i-2) \binom{n}{i} + 6 \cdot \sum_{i=\lceil \frac{n}{2} \rceil+1}^{n} (i-3) \binom{n}{i} \leq (5n-21) \cdot 2^{n-1} \qquad (4.31)$$

$$4 \cdot \sum_{i=0}^{\lceil \frac{n}{2} \rceil} (i-2) \binom{n}{i} + 8 \cdot \sum_{i=\lceil \frac{n}{2} \rceil+1}^{n} (i-4) \binom{n}{i} \leq (6n-28) \cdot 2^{n-1} \qquad (4.32)$$

Proof. We consider the cases when n is even and n is odd.
n **is even:** we have

$$\lceil \tfrac{n}{2} \rceil = \frac{n}{2}$$

$$\sum_{i=0}^{\frac{n}{2}} \binom{n}{i} = 2^{n-1} + \frac{1}{2} \binom{n}{\frac{n}{2}}$$

$$\sum_{i=\frac{n}{2}+1}^{n} \binom{n}{i} = 2^{n-1} - \frac{1}{2} \binom{n}{\frac{n}{2}}$$

Then

$$4 \cdot \sum_{i=1}^{\frac{n}{2}} (i-2) \binom{n}{i} + 8 \cdot \sum_{i=\frac{n}{2}+1}^{n} (i-3) \binom{n}{i}$$

$$= 4 \cdot \sum_{i=1}^{\frac{n}{2}} i \binom{n}{i} - 8 \cdot \sum_{i=1}^{\frac{n}{2}} \binom{n}{i} + 8 \cdot \sum_{i=\frac{n}{2}+1}^{n} i \binom{n}{i} - 24 \cdot \sum_{i=\frac{n}{2}+1}^{n} \binom{n}{i}$$

$$\stackrel{\text{Lemma 4.4}}{=} 4n \cdot 2^{n-2} - 8 \left(2^{n-1} + \frac{1}{2} \binom{n}{\frac{n}{2}} \right) + 8n \cdot 2^{n-2} - 24 \left(2^{n-1} - \frac{1}{2} \binom{n}{\frac{n}{2}} \right)$$

$$= (6n - 32) \cdot 2^{n-1} + 8 \binom{\frac{n}{2}}{n}$$

For $n \geq 1$, we have

$$\binom{\frac{n}{2}}{n} \leq 2^{n-1}$$

Hence

$$4 \cdot \sum_{i=1}^{\frac{n}{2}} (i-2) \binom{n}{i} + 8 \cdot \sum_{i=\frac{n}{2}+1}^{n} (i-3) \binom{n}{i}$$

$$\leq (6n - 32) \cdot 2^{n-1} + 8.2^{n-1} = (6n - 24) \cdot 2^n$$

n **is odd:** we have

$$\left\lceil \frac{n}{2} \right\rceil = \frac{n+1}{2}$$

$$\sum_{i=0}^{\frac{n+1}{2}} \binom{n}{i} = 2^{n-1} + \binom{\frac{n+1}{2}}{n}$$

$$\sum_{i=\frac{n+3}{2}}^{n} \binom{n}{i} = 2^{n-1} - \binom{\frac{n+1}{2}}{n}$$

Then

$$
4 \cdot \sum_{i=1}^{\frac{n+1}{2}} (i-2)\binom{n}{i} + 8 \cdot \sum_{i=\frac{n+3}{2}}^{n} (i-3)\binom{n}{i}
$$

$$
= 4 \cdot \sum_{i=1}^{\frac{n+1}{2}} i\binom{n}{i} - 8 \cdot \sum_{i=1}^{\frac{n+1}{2}} \binom{n}{i} + 8 \cdot \sum_{i=\frac{n+3}{2}}^{n} i\binom{n}{i} - 24 \cdot \sum_{i=\frac{n+3}{2}}^{n} \binom{n}{i}
$$

$$
\overset{\text{Lemma 4.4}}{=} 4\left(n \cdot 2^{n-2} + \frac{n+1}{4}\binom{\frac{n+1}{2}}{n}\right) - 8\left(2^{n-1} + \binom{\frac{n+1}{2}}{n}\right)
$$

$$
+ 8\left(n \cdot 2^{n-2} - \frac{n+1}{4}\binom{\frac{n+1}{2}}{n}\right) - 24\left(2^{n-1} - \binom{\frac{n+1}{2}}{n}\right)
$$

$$
= (6n - 32) \cdot 2^{n-1} - (n-15)\binom{\frac{n+1}{2}}{n}
$$

Since $\binom{\frac{n+1}{2}}{n} \leq 3 \cdot 2^{n-3}$, we get

$$
4 \cdot \sum_{i=1}^{\frac{n+1}{2}} (i-2)\binom{n}{i} + 8 \cdot \sum_{i=\frac{n+3}{2}}^{n} (i-3)\binom{n}{i}
$$

$$
\leq (6n - 32) \cdot 2^{n-1} - (n-15) \cdot 3 \cdot 2^{n-1}
$$

$$
= (6n - 24) \cdot 2^{n-1} - 8 \cdot 2^{n-1} - (n-15) \cdot 3 \cdot 2^{n-3}
$$

$$
= (6n - 24) \cdot 2^{n-1} - (3n-13) \cdot 2^{n-3}
$$

$$
\leq (6n - 24) \cdot 2^{n-1}.
$$

The other inequations are obtained using the same argument. ∎

Theorem 4.2. *An n-variable single-target gate with $n \geq 6$ can be realized with at most*

- $us_{\text{NCT}_{\text{B1}}}(n) \leq (6n - 30) \cdot 2^{n-2}$
- $us_{\text{NCT}_{\text{NC}}}(n) \leq (5n - 26) \cdot 2^{n-2}$
- $us_{\text{NCT}_{\text{MI}}}(n) \leq (6n - 34) \cdot 2^{n-2}$

NCT gates.

Proof. An n-variable single-target gate can be realized with at most $us_{\mathrm{MCT}}(n) = 2^{n-1}$ MCT gates. This follows from the PPRM representation, which is canonical for a given function when disregarding the order of the product terms. Hence, there exists a control function $g \in \mathscr{B}_{n-1}$ for which the PPRM expression consists of all 2^{n-1} product terms, and therefore

$$us_{\mathrm{MCT}}(n) \leq 2^{n-1} \leq \sum_{i=0}^{n-1} \binom{n-1}{i} \tag{4.33}$$

with $\binom{n-1}{i}$ being the total number of product terms that have i literals, i.e., the total number of Toffoli gates that have i controls. Consider now the number of gates after mapping to the NCT gate library: in [83], it has been shown that each MPMCT gate, with i controls and $i \leq \lceil \frac{n-1}{2} \rceil$, can be decomposed to $4(i-2)$ Toffoli gates. Otherwise, it can be decomposed to a circuit using one of the mapping algorithms that requires one ancilla.

When the B1 mapping is chosen as a one ancilla mapping algorithm, each MPMCT gate, with i controls and $i > \lceil \frac{n-1}{2} \rceil$ can be decomposed to $8(i-3)$ Toffoli gates. We thus have

$$us_{\mathrm{NCT_{B1}}}(n) \leq 4 \cdot \sum_{i=0}^{\lceil \frac{n-1}{2} \rceil} (i-2)\binom{n-1}{i} + 8 \cdot \sum_{i=\lceil \frac{n-1}{2} \rceil + 1}^{n-1} (i-3)\binom{n-1}{i}$$

$$\overset{\text{Corollary 4.1}}{\leq} (6(n-1) - 24) \cdot 2^{(n-1)-1}$$

$$\leq (6n - 30) \cdot 2^{n-2}.$$

When the NC mapping is applied, to decompose the i-control MPMCT gates with $i > \lceil \frac{n-1}{2} \rceil$, then each gate can be mapped to $6(i-3)$ Toffoli gates. Therefore

$$us_{\mathrm{NCT_{B1}}}(n) \leq 4 \cdot \sum_{i=0}^{\lceil \frac{n-1}{2} \rceil} (i-2)\binom{n-1}{i} + 6 \cdot \sum_{i=\lceil \frac{n-1}{2} \rceil + 1}^{n-1} (i-3)\binom{n-1}{i}$$

$$\overset{\text{Corollary 4.1}}{\leq} (5(n-1) - 21) \cdot 2^{(n-1)-1}$$

$$\leq (5n - 26) \cdot 2^{n-2}.$$

When the MI mapping is adopted, each i-control MPMCT gate with $i > \lceil \frac{n-1}{2} \rceil$ can be decomposed to $8(i-4)$ Toffoli gates. We thus have

$$us_{\text{NCT}_{\text{B1}}}(n) \leq 4 \cdot \sum_{i=0}^{\lceil \frac{n-1}{2} \rceil} (i-2) \binom{n-1}{i} + 8 \cdot \sum_{i=\lceil \frac{n-1}{2} \rceil+1}^{n-1} (i-4) \binom{n-1}{i}$$

$$\overset{\text{Corollary 4.1}}{\leq} (6(n-1) - 28) \cdot 2^{(n-1)-1}$$

$$\leq (6n - 34) \cdot 2^{n-2}.$$

∎

4.4.2.2 Upper Bounds Based on General ESOP Expressions

Even tighter bounds can be obtained when using general ESOP expressions instead of PPRM expressions. Note that this includes the consideration of negative control lines.

Theorem 4.3. *An n-variable single-target gate can be realized, if $n \geq 6$, with at most*

- $us_{\text{NCT}_{\text{B1}}}(n) \leq 29(n-4) \cdot 2^{n-5}$
- $us_{\text{NCT}_{\text{NC}}}(n) \leq \frac{87}{4}(n-4) \cdot 2^{n-5}$
- $us_{\text{NCT}_{\text{MI}}}(n) \leq 29(n-5) \cdot 2^{n-5}$

NCT gates.

Proof. The best known upper bound on the number of product terms in a minimum ESOP form for an n-variables Boolean function is [50]

$$29 \cdot 2^{n-7} \quad \text{with} \quad n \geq 7. \tag{4.34}$$

Hence, the ESOP expression of the control function $g \in \mathscr{B}_{n-1}$ consists of at most $29 \cdot 2^{n-8}$ product terms. Each product term has at most $n-1$ terms in the worst case, and so mapping a product term (i.e., MPMCT gate with at most $n-1$ controls) using the B1 mapping gives at most $8(n-4)$ Toffoli gates. Hence, at most $29(n-4) \cdot 2^{n-5}$ Toffoli gates are required when the B1 mapping is applied. Therefore, we have $us_{\text{NCT}_{\text{B1}}}(n) \leq 29(n-4) \cdot 2^{n-5}$.

The remaining upper bounds may be observed using the same argument. ∎

4.4.2.3 Upper Bounds Based on Functional Decomposition

In this section, we derive upper bounds based on functional decomposition in an induction proof. We use exact bounds for the base case that were found using exhaustive search in combination with optimal synthesis [54].

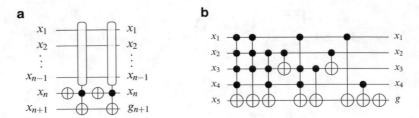

Fig. 4.20 ST gate decomposition based on MCT gates. (**a**) ST gate decomposition. (**b**) MCT circuit

Theorem 4.4. *For an n-variable single-target gate we have for $n \geq 6$:*

- $us_{NCT_{BI}}(n) \leq (5n - 31) \cdot 2^{n-1} + (2^{n-4} - 2)$
- $us_{NCT_{NC}}(n) \leq (15n - 93) \cdot 2^{n-3} + (2^{n-4} - 2)$
- $us_{NCT_{MI}}(n) \leq (5n - 36) \cdot 2^{n-1} + (2^{n-4} - 2)$

NCT gates, where the first term in each bound refers to the number of Toffoli gates and the second term refers to the number of NOT gates.

Proof. We first show an upper bound of $us_{MCT}(n) = 2^{n-5} \cdot 10 + 2^{n-4} - 2$ for the number of MCT gates if $n \geq 5$. The proof is obtained by induction on n. For the base case let $n = 5$. Using exhaustive search we enumerated all 65,536 Boolean functions over four variables that can be represented by a 5-bit single-target gate and for each one we obtain the minimal circuit using an exact synthesis approach [54]. The largest circuit required ten MCT gates as depicted in Fig. 4.20b:

$$us_{MCT}(5) = 1 + 3 + 0 + 5 + 1 = 10, \qquad (4.35)$$

where the terms partition the Toffoli gates by their number of controls from 4 controls to 0 controls.

In the induction step, let's consider an $(n + 1)$-bit single-target gate. By applying the Shannon decomposition the gate can be decomposed into two n-bit single-target gates and 2 NOT gates (see Fig. 4.20a), we obtain

$$
\begin{aligned}
us_{MCT}(n + 1) &= 2 \cdot us_{MCT}(n) + 2 \\
&= 2(2^{n-5} \cdot 10 + 2^{n-4} - 2) + 2 \\
&= 2^{n-4} \cdot 10 + 2^{n-3} - 4 + 2 \\
&= 2^{(n+1)-5} \cdot 10 + 2^{(n+1)-4} - 2
\end{aligned}
$$

Hence, a single-target gate on n lines requires at most $2^{n-5} \cdot 10 + 2^{n-4} - 2$ MCT gates.

For the mapping to NCT gates we need to take a closer look at the number of controls for each MCT gate. According to (4.35), we have the following distribution for each n-bit single-target gate:

$$
\begin{array}{ccccccc}
\text{\#controls} & n-1 & n-2 & n-3 & n-4 & n-5 & 0 \\
\text{\#gates} & 2^{n-5} & 3 \cdot 2^{n-5} & 0 & 5 \cdot 2^{n-5} & 2^{n-5} & 2^{n-4}-2
\end{array}
\tag{4.36}
$$

As an example, consider now the B1 mapping which requires $8(n-3)$ NCT gates for an MCT gate with n controls. Therefore:

$$
\begin{aligned}
us_{\text{NCT}_{\text{B1}}}(n) &= 2^{n-5}(8(n-4) + 3 \cdot 8(n-5) + 5 \cdot 8(n-7) + 8(n-8)) + (2^{n-4} - 2) \\
&= 2^{n-2}((n-4) + 3(n-5) + 5(n-7) + (n-8)) + (2^{n-4} - 2) \\
&= 2^{n-1}(5n - 31) + (2^{n-4} - 2)
\end{aligned}
$$

The bounds based on the other mappings can be derived analogously. ∎

Better bounds can be found when extending the decomposition-based approach for MPMCT gates.

Theorem 4.5. *An n-variable single-target gate can be realized, if $n \geq 6$, with at most*

- $us_{\text{NCT}_{\text{B1}}}(n) \leq (3n - 16) \cdot 2^{n-1}$
- $us_{\text{NCT}_{\text{NC}}}(n) \leq (\frac{9}{4}n - 12) \cdot 2^{n-1}$
- $us_{\text{NCT}_{\text{MI}}}(n) \leq (3n - 19) \cdot 2^{n-1}$

NCT gates.

Proof. We are using the same idea for the proof as for Theorem 4.4. The largest minimal circuit contains six MPMCT gates and is shown in Fig. 4.21b. Together with the decomposition depicted in Fig. 4.21a an upper bound for the number of gates $us_{\text{MPMCT}}(n) = 2^{n-5} \cdot 6$ is obtained for $n \geq 6$.

As an example, consider again the B1 mapping which requires $8(n - 3)$ NCT gates for an MPMCT gate with n controls. Based on the new bound for MPMCT

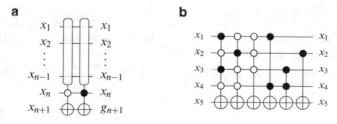

Fig. 4.21 ST gate decomposition based on MPMCT gates. (a) ST gate decomposition. (b) MPMCT circuit

gates and the following distribution for each n-bit single-target gate:

$$
\begin{array}{c|ccccc}
\#\text{controls} & n-1 & n-2 & n-3 & n-4 & n-5 \\
\#\text{gates} & 2 \cdot 2^{n-5} & 2^{n-5} & 2 \cdot 2^{n-5} & 2^{n-5} & 0
\end{array}
\tag{4.37}
$$

We have

$$
\begin{aligned}
us_{\text{NCT}_{\text{B1}}}(n) &\leq 2^{n-5}(2 \cdot 8(n-4) + 8(n-5) + 2 \cdot 8(n-6) + 8(n-7)) \\
&= (6n - 32)2^{n-2} \\
&= (3n - 16)2^{n-1}
\end{aligned}
$$

NCT gates. The bounds based on the other mappings can be derived analogously.   ∎

Comparing the upper bounds for single-target gates based on ESOPs and function decomposition, we observe that general PPRM expressions give the best upper bounds. In fact, it gives the exact number of MCT gates and their exact number of controls. In the remainder of this book, we will focus on the upper bounds that are derived from PPRM expressions.

4.4.3 Upper Bounds for NCT Circuits

In the following, we give new upper bounds on the size of NCT circuits for general reversible functions using the transformation based synthesis approach with different mapping strategies:

Theorem 4.6. *Using the transformation based synthesis algorithm, an NCT based circuit has at most*

- $uc_{\text{NCT}_{\text{B1}}}(n) \leq (10n - 28) \cdot 2^{n-1}$
- $uc_{\text{NCT}_{\text{NC}}}(n) \leq (9n - 25) \cdot 2^{n-1}$
- $uc_{\text{NCT}_{\text{MI}}}(n) \leq (10n - 32) \cdot 2^{n-1}$

NCT gates.

Proof. A reversible circuit synthesized with the transformation based approach has at most n NOT gates, $(n-1) \cdot 2^{n+1} - n^2 + 4$ CNOT gates, and $\sum_{i=2}^{n-1} \binom{n}{i}$ MCT gates where i denotes the number of controls on each gate.

Based on the B1 mapping, after mapping the $\sum_{i=2}^{n-1} \binom{n}{i}$ MCT gates, we get

$$
4 \cdot \sum_{i=2}^{\left\lceil \frac{n}{2} \right\rceil}(i-2)\binom{n}{i} + 8 \cdot \sum_{i=\left\lceil \frac{n}{2} \right\rceil + 1}^{n-1}(i-3)\binom{n}{i} \text{ NCT gates. Hence,}
$$

$$uc_{NCT_{B1}}(n) \leq n + (n-1) \cdot 2^{n+1} - n^2 + 4 + 4 \cdot \sum_{i=2}^{\lceil \frac{n}{2} \rceil}(i-2)\binom{n}{i}$$

$$+ 8 \cdot \sum_{i=\lceil \frac{n}{2} \rceil+1}^{n-1}(i-3)\binom{n}{i}$$

$$\overset{Corollary\ 4.1}{\leq} \quad n + (n-1) \cdot 2^{n+1} - n^2 + 4 + (6n-24) \cdot 2^{n-1}$$

$$\leq (10n-28) \cdot 2^{n-1} + n - n^2 + 4 \leq (10n-28) \cdot 2^{n-1}$$

Using the NC mapping,

$$uc_{NCT_{NC}}(n) \leq n + (n-1) \cdot 2^{n+1} - n^2 + 4 + 4 \cdot \sum_{i=2}^{\lceil \frac{n}{2} \rceil}(i-2)\binom{n}{i}$$

$$+ 6 \cdot \sum_{i=\lceil \frac{n}{2} \rceil+1}^{n-1}(i-3)\binom{n}{i}$$

$$\overset{Corollary\ 4.1}{\leq} \quad n + (n-1) \cdot 2^{n+1} - n^2 + 4 + (5n-21) \cdot 2^{n-1}$$

$$\leq (9n-25) \cdot 2^{n-1} + n - n^2 + 4$$

$$\leq (9n-25) \cdot 2^{n-1} + n - n^2 + 4$$

$$\leq (9n-25) \cdot 2^{n-1}$$

Using the MI mapping,

$$uc_{NCT_{MI}}(n) \leq n + (n-1) \cdot 2^{n+1} - n^2 + 4 + 4 \cdot \sum_{i=2}^{\lceil \frac{n}{2} \rceil}(i-2)\binom{n}{i}$$

$$+ 8 \cdot \sum_{i=\lceil \frac{n}{2} \rceil+1}^{n-1}(i-4)\binom{n}{i}$$

$$\overset{Corollary\ 4.1}{\leq} \quad n + (n-1) \cdot 2^{n+1} - n^2 + 4 + (6n-28) \cdot 2^{n-1}$$

$$\leq (10n-32) \cdot 2^{n-1} + n - n^2 + 4$$

$$\leq (10n-32) \cdot 2^{n-1}$$

∎

4.5   Summary

In this chapter, we first proposed a mapping approach that starts with single-target gates and therefore differs from the standard mapping approach that has been known as the state-of-the-art for the last two decades. We observed that incorporating Boolean decomposition in the mapping process of single-target gates often leads to better quantum realizations. Motivated by this, we introduced an improved mapping scheme which uses a constant number of ancillas and exploits the Boolean decomposition when generating the quantum gate circuits for a given single-target gate. Our approach results in quantum circuits with a smaller NCV quantum cost as well as a lower T-depth cost compared to the results achieved by the standard mapping. Experiments show that considering the Boolean decomposition on the mapping of ST gates leads to a quantum cost reduction of 20 % on average compared to standard mapping scheme.

Then, we extended and improved the existing mapping algorithms of reversible circuits into quantum circuits using the Clifford $+ T$ quantum library. To the best of our knowledge, no mapping algorithm has been presented for the Clifford $+ T$ gate library so far. Our novel mapping approaches lead to smaller T-depth reduced by more than 50 % compared to previous work as shown in the experimental results. In particular, we have shown that the NC together with B1 mapping algorithm is the best mapping strategy with respect to T-depth.

Finally, we studied the complexity of NCT circuits obtained after applying the B1, B2, NC, and MI mapping algorithms. We gave the upper bounds on the number of NCT gates required to realize MPMCT gate, ST gate, and general reversible circuits with respect to each mapping approach. We proved that reversible function over n variables can be realized with at most $(10n - 28) \cdot 2^{n-1}$, $(9n - 25) \cdot 2^{n-1}$, or $(10n - 32) \cdot 2^{n-1}$ NCT gates based on the mapping approach B1, NC, or MI, respectively.

Chapter 5
Optimizations and Complexity Analysis on the Quantum Level

After mapping a reversible circuit into a functionally equivalent quantum circuit, the resulting circuit is not optimal with respect to considered cost metrics due to the existing non-optimal synthesis as well as mapping approaches, particularly for large circuits. Moreover, physical developments for emerging technologies constantly lead to new constraints, e.g., the depth and the NNC. For that reason, post-mapping optimizations are applied to optimize the value of the cost metrics with respect to the used library. In particular, many existing optimization methods target the reduction of the quantum cost and the depth [83, 128] of quantum circuits. In the following section, related work is summarized. Afterwards, a depth optimization approach is given in Sect. 5.2. Then, in Sect. 5.3 an algorithm for reducing the quantum cost is explained. Finally a study of the complexity of quantum implementations is given in Sect. 5.4.

5.1 Related Work

We review the existing optimization algorithms targeting the quantum cost and the depth. Then, we briefly present the complexity of quantum circuits implementing reversible functions.

5.1.1 Optimization of Quantum Circuits

5.1.1.1 NCV-Cost Reduction

There are many post-mapping optimization schemes that aim at reducing the NCV-cost for circuits based on the *NCV* library. These techniques attempt to apply templates or reduction rules by deleting identical gates or replacing circuits of gates

© Springer International Publishing Switzerland 2016
N. Abdessaied, R. Drechsler, *Reversible and Quantum Circuits*,
DOI 10.1007/978-3-319-31937-7_5

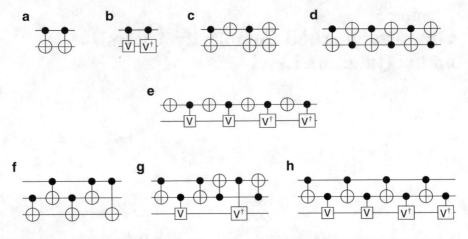

Fig. 5.1 Quantum templates with 2 or 3 lines. (**a**) Template 1. (**b**) Template 2. (**c**) Template 3. (**d**) Template 4. (**e**) Template 5. (**f**) Template 6. (**g**) Template 7. (**h**) Template 8

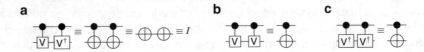

Fig. 5.2 Reduction rules for the NCV circuits. (**a**) Rule 1. (**b**) Rule 2. (**c**) Rule 3

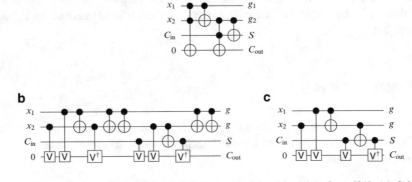

Fig. 5.3 Simplification of an NCV circuit for the 1-bit full adder taken from [81]. (**a**) Original circuit. (**b**) Mapped circuit. (**c**) Optimized circuit

with smaller ones. For this purpose, the gates are rearranged together to match the reduction rules by means of moving rules. For example, the quantum template matching algorithms that are proposed in [80, 83, 103, 104, 106] employ the classical moving rules (CMR) defined in Sect. 3.1.1.1. Figures 5.1 and 5.2 show a set of templates and several reduction rules for optimizing NCV circuits.

Example 5.1. A quantum circuit for the 1-bit full adder can be constructed from the reversible implementation shown in Fig. 5.3a. First, each gate is substituted with its

Fig. 5.4 LLP moving rule

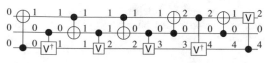

Fig. 5.5 DDMF moving rule

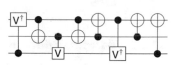

quantum realization. After applying the template matching algorithm, gates 5 and 6 match the template given in Fig. 5.1a, similarly for gates 11 and 12. Furthermore, gates 4 and 6 can be moved together and match the template outlined in Fig. 5.1b. Therefore, six gates are deleted leading to the circuit shown in Fig. 5.3c.

Also, different rule-based algorithms have been presented [86, 111, 128]. In particular, the approach given in [112] has proven its efficiency due to the use of better gate movement properties. Indeed, the moving rules based on LLP have given further possible gate rearrangement in comparison with the CMR. However, this procedure does not find all the line segments that have the same functionality.

Example 5.2. As depicted in Fig. 5.4, the circuit between the second gate and the ninth gate present an identity circuit, i.e., the first gate can be moved to last position. Accordingly, the labels on the input side of the second gate and the output side of the ninth gate should be the same. Besides, the third line between the circuit from the position two to the position nine should not contain any control, which is not true in this case. Thus, taking into account the LLP constraints, the two gates could not be moved together although it is possible.

To overcome this problem, Sasanian et al. in [112] propose the *Decision Diagrams for Matrix Functions* (DDMF) data structure instead of LLP to represent functions on circuit line segments. Using the DDMF allows to find all identity segments and gives additional gate reduction. The limitation of this representation is that it is only applicable to reversible circuits and a certain class of quantum circuits called *Semi-Classical Quantum Circuits* (SCQC). For SCQC, entangled gates should be removed and replaced by SCQC sub-circuits. Removing the entanglement adds extra quantum costs that are sometimes comparable to the improvements achieved by the LLP procedure.

Example 5.3. Even though the first gate in Fig. 5.5 could be moved next to the last gate since the cascade from the second till the seventh gate is the identity circuit, but the DDMF approach cannot be applied to this circuit because it contains an entangled state generated from the first gate.

Fig. 5.6 An optimal T-depth implementation of the 1-bit full adder

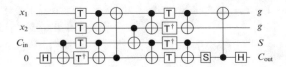

5.1.1.2 T-Depth Reduction

The optimization methods for the Clifford $+\,T$ circuits mainly focus on reducing the number of T gates and the T-*depth* of the resulting circuits. This is because fault-tolerant implementations of T gates are considerably more expensive than those of the Clifford gates [11, 48]. Thus, several algorithms have been proposed to minimize the T-*count* [53] and the T-*depth* [10, 11, 120].

The algorithm presented in [53] describes a method that performs an exhaustive search for a circuit that implements an n-qubit unitary matrix U by use of the minimal number of T gates. Similarly, the work introduced in [10] addressed the exact optimization of T-*depth* for small circuits composed of four qubits at maximum. This is done by applying an exhaustive search algorithm to find the optimal T-*depth* realization.

Example 5.4. Figure 5.6 depicts a quantum circuit based on the Clifford $+\,T$ library and taken from [10]. This circuit is an optimal T-depth implementation for the 1-bit full adder shown in Fig. 5.3a. It is derived using the T-depth optimization algorithm [10] mentioned above.

Another approach proposed an algorithm for reducing the T-*depth* of quantum circuits over the gate library *{CNOT, T}* [11]; that is, the Clifford $+\,T$ library without the Hadamard gate. The algorithm deletes redundant T gates by computing the total phase and parallelizing the T gates through *matroid partitioning*. The idea is based on decomposing a given function into minimal number of linear Boolean functions. Then, re-synthesizing each one with an optimized T-*depth* realization. Later in [11], this algorithm is extended for circuits built with the Clifford $+\,T$ library. In this case, the same approach is applied to the subcircuits between the H gates, afterwards an optimization process is applied which detects the identical gates and deletes them.

In [120], a class of circuits were proposed where the T-depth can be reduced to 1 by using a sufficient number of ancillae.

Example 5.5. Figure 5.7 shows a circuit that realizes a 1-bit full adder using four additional ancillae set to 0. The circuit has a T-depth of 1.

5.1.1.3 Depth Reduction

Beyond the quantum cost, the depth of a quantum circuit is important. Although several synthesis approaches that consider depth have recently been introduced (cf. [13, 14, 20, 83]), the majority of design methods do not consider this metric.

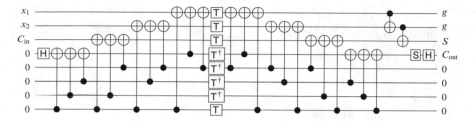

Fig. 5.7 A T-depth 1 implementation of a 1-bit full adder

As an example, in [13, 14] a cycle representation is chosen and input cycles are partitioned into three subsets. Each subset is synthesized independently on a different set of ancillae (initialized to zero) in parallel. This method requires $2n$ additional lines and focuses only on reducing the depth of reversible circuits rather than the quantum circuits. This is crucial since the execution times for two reversible gates can differ significantly when the respective quantum circuit mapping is taken into account. Another post-synthesis approach has been presented in [83] that makes use of a special class of templates to reduce the depth of a given NCV circuit.

5.1.2 Complexity of Quantum Circuits

Many studies have focused on the upper bound of the number of elementary quantum gates in quantum circuits. The transformation based algorithm as mentioned in [82] leads to quantum circuits over the NCV library with an NCV-cost of at most:

$$11n \cdot 2^n + o(n \cdot 2^n) \tag{5.1}$$

Comparatively, in [108] the authors have confirmed that using a cycle based synthesis approach, any reversible function over n variable can be realized with at most

$$8.5n \cdot 2^n \tag{5.2}$$

NCV gates.

5.2 Depth Optimization for NCV Circuits

In this section, we present an idea on how depth of quantum circuits can be reduced by adding an ancilla to the circuit. Based on this idea, two depth-optimization approaches are presented. The first method aims at reducing the depth by applying the reduction gate-per-gate, whereas the second method focuses on the whole

---

Algorithm 5: Circuit depth

Input: Quantum circuit $G = T_1(C_1, t_1) \cdots T_k(C_k, t_k)$
Output: Depth d
// Initialize
1 $d \leftarrow 1$
2 $i \leftarrow 1$
3 $j \leftarrow 1$
4 **for** $j \leftarrow 1$ **to** Lines (G) **do**
5 | $b_j \leftarrow 0$
6 **end**
7 **while** $i < k$ **do** // Terminate?
8 | **foreach** $j \in C_i \cup \{t_i\}$ **do** $b_j \leftarrow b_j + 1$ // Apply gate
9 | $j \leftarrow 0$
10 | **while** $j <$ Lines (G) *and* $b_j \neq 2$ **do** // Gates do not overlap?
11 | | $j \leftarrow j + 1$
12 | **end**
13 | **if** $j \neq k$ **then** // When gates overlap
14 | | **for** $j \leftarrow 0$ **to** Lines (G) **do**
15 | | | **if** $j \in C_i \cup \{t_i\}$ **then**
16 | | | | $b_j \leftarrow 1$
17 | | | **else**
18 | | | | $b_j \leftarrow 0$
19 | | | **end**
20 | | | $d \leftarrow d + 1$
21 | | **end**
22 | **end**
23 | $i \leftarrow i + 1$
24 **end**

---

circuit. Our optimization techniques enable a concurrent application of gates which significantly reduces the depth of the circuit. Experiments show that reductions of 28 % on average can be achieved by the proposed schemes compared to the original depth of given circuits. Both methods have been published in [2].

Algorithm 5 computes a circuit depth d. Given a quantum circuit

$$G = g_1(C_1, t_1) \ldots g_k(C_k, t_k)$$

over n variables $x_1, \ldots, x_n$. The algorithm determines the depth d of the circuit according to Definition 2.21 (see page 35) by applying a greedy search to gates that can be executed in parallel. For the computation, we are making use of the integers $b_1, \ldots, b_n$.

Example 5.6. Figure 5.8c illustrates the depth for the reversible circuit shown in Fig. 5.8b.

Although the *coherence time* (the time a qubit can keep its quantum state) and the *gate operation time* (the time a gate needs to perform its operation) may vary from one technology to another (cf. Table III, [84]), keeping the overall execution

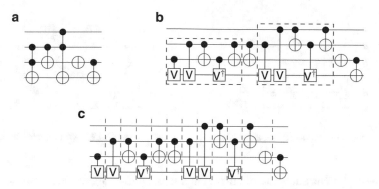

Fig. 5.8 Quantum depth. (**a**) Reversible circuit. (**b**) Quantum circuit. (**c**) Depth computing

time as small as possible is essential in all these cases. Consequently, the depth metric can be applied in a generic manner, as it provides a proper model which can be considered already at the synthesis stage in the absence of precise technological constraints. Motivated by this, we investigate in the rest of this section how the depth of a quantum circuit can be reduced. The remainder of this section is structured as follows. The general idea is presented in Sect. 5.2.1. Afterwards, both proposed approaches are described and evaluated in Sects. 5.2.2 and 5.2.3, respectively.

5.2.1 General Idea

Keeping the number of circuit lines as small as possible is required in the synthesis of quantum circuits. This is mainly motivated by the fact that each circuit line has to be represented by a qubit, which is a very limited resource. Nevertheless, evaluations also showed that a (slight) extension of a circuit with additional lines may have significant benefits. For example in Sect. 4.1.1, it has been demonstrated that a larger number of circuit lines allow a much cheaper mapping of reversible circuits to quantum circuits in terms of gate count. In [145], evaluations showed that using twice the number of circuit lines reduces the NCV-cost by up to two orders of magnitude. Eventually, this led to a post-synthesis optimization approach [90] which enables reductions in NCV-cost of up to 69 % only by adding a single additional line to the circuit.

In this section, we show that similar concepts also help in reducing the depth of quantum circuits. We are following the established design flow reviewed in Chap. 1, i.e., first a reversible circuit is realized which afterwards is mapped to a quantum circuit. Indeed, by incorporating an additional ancilla during this process makes a depth-aware optimization possible.

Following the concept from [90], ancillae (initialized to 0) can now be applied in order to "buffer" values of circuit lines so that they can be reused later by other gates.

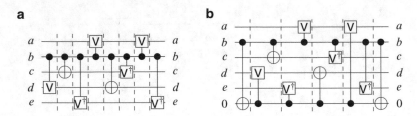

Fig. 5.9 Depth reduction using an ancilla. (**a**) Initial circuit. (**b**) Circuit with reduced depth

Whenever the current value of an ancilla h is 0, another signal line x can be copied to h by appending a *copy gate* $C(\{x\}, h)$ to the circuit. The ancilla can be restored with the same gate if no other gate has used h as target line in between.

In [90], this buffering has been exploited to remove common control lines connections between Toffoli gates in order to reduce the NCV-cost. However, the same concept can be similarly applied to reduce the depth of quantum circuits as illustrated by the following example.

Example 5.7. Figure 5.9a shows a circuit in which no gate can be performed in parallel since all the gates share the same control line b. In Fig. 5.9b, an ancilla has been added to copy the value of b. By performing this, the gates can be rearranged which reduces the depth from 8 to 6.

Clearly, Example 5.7 presents a rather artificial circuit. However, based on this general idea we are proposing different optimization approaches whose evaluations show that indeed a significant reduction of depth in quantum circuits can be achieved.

5.2.2 Optimization Approaches

Motivated by the general idea outlined above, two optimization approaches are proposed in this section which aim at reducing the depth by exploiting additional circuit lines. The first approach follows a local scheme, i.e., considering each Toffoli gate independently, while the second approach considers the whole circuit instead.

5.2.2.1 Consideration of Single Gates

The availability of an ancilla as introduced in the previous section allows for an improvement of the mapping scheme reviewed in Sect. 4.1.1. Recall that, using the default mapping scheme, each Toffoli gate is mapped to a quantum realization of depth 5 as shown in Fig. 5.10b. However, as the second and the third gate share the same control line, an additional ancilla allows for a concurrent execution of

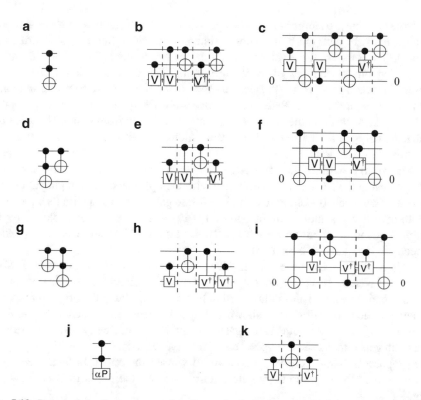

Fig. 5.10 Proposed depth-aware mapping for each single gate. (**a**) Toffoli gate. (**b**) Original mapping. (**c**) Proposed mapping. (**d**) Peres gate. (**e**) Original mapping. (**f**) Proposed mapping. (**g**) Inverse Peres gate. (**h**) Original mapping. (**i**) Proposed mapping. (**j**) αP gate. (**k**) Original mapping

Fig. 5.11 Application of the local scheme to the circuit from Fig. 5.8a

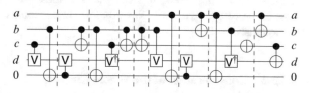

both gates as shown in Fig. 5.10c. Additionally, since the copy gates can be inserted without increasing the depth, a depth reduction for the quantum circuit realization for each Toffoli gate from 5 to 4 can be obtained.

Example 5.8. Consider again the reversible circuit from Fig. 5.8a. Using the established mapping scheme from Sect. 4.1.1, the resulting quantum circuit has a depth of 12 (as shown in Fig. 5.8b; none of the gates except for the single NOT gate can be executed concurrently). In contrast, applying the additional ancilla as proposed in Fig. 5.10b, the circuit depicted in Fig. 5.11 results. This reduces the depth from 12 to 9.

Note that this procedure can also be applied to MPMCT gates with more than two control lines. In fact, the best mapping schemes for NCV library (as described in Sect. 4.1) decompose these gates using either the NC or the MS mapping algorithm depending on the number of available ancilla. The obtained circuit consists of Peres, inverse Peres, and αP gates. In this case, the depth-optimized mapping consists of replacing each Peres and its inverse to the quantum gates as proposed in Fig. 5.10f, i such that the depth of both gates is reduced from 4 to 3. Note that αP gates are mapped to their equivalent quantum circuits without applying any depth-optimization technique.

Algorithm 6 outlines the approach explained above. First, a reversible circuit is mapped using the NC mapping combined with the optimized B2 mapping approach (line 3). Then each Toffoli, Peres, and its inverse gate is replaced with its equivalent depth-optimized quantum circuits (lines 11–17). Each αP gate is replaced by its original quantum circuit with (line 18). Note that each NOT and CNOT is pended directly to the quantum circuit.

This scheme is not beneficial in all cases. In fact, if two concurrent Toffoli, Peres, or Inverse Peres gates are mapped to a quantum circuit, the original mapping leads to better results. This is illustrated in Fig. 5.12. Applying the original mapping scheme to the two Toffoli gates shown in Fig. 5.12a leads to the quantum circuit as shown in Fig. 5.12b. As both Toffoli gates are applied concurrently, also the resulting quantum gate circuits can be applied concurrently, i.e., a depth of 5 results. Applying the proposed scheme from Fig. 5.10b would worsen the result. In fact, the ancilla together with the required copy gates would increase the depth to 7 as shown in Fig. 5.12c.

Consequently, this scheme is only applied in cases where an actual depth improvement can be achieved. However, experiments summarized in Sect. 5.2.3 clearly confirm that substantial improvements with respect to the depth can be achieved. As a drawback, this obviously comes with the price of increased NCV-cost in the resulting circuit. Nevertheless, experiments show the resulting increase is moderate.

5.2.2.2 Consideration of the Whole Circuit

While so far the ancilla has been exploited in a local context, also a global consideration turns out to be beneficial. After mapping a circuit to its NCV quantum equivalent circuit, the idea is to identify subcircuits of gates sharing the same control line (see line 4 in Algorithm 7) and use the ancilla in order to partition the gates. Then, each consecutive pair of gates in such a circuit can be concurrently executed by using the original control line for the first gate and the copied value at the ancilla for the second gate (see lines 8–10 in Algorithm 7).

Example 5.9. Figure 5.13a shows a quantum circuit composed of gates that share the same control lines. Using the ancilla, an equivalent realization as shown in Fig. 5.13b can be derived. This reduces the depth from 5 to 4.

Algorithm 6: Local depth optimization

---

1 **Input:** Reversible circuit G
   Output: Local Depth optimized quantum circuit Q
   `// Initialize`
2 $i \leftarrow 1$
   `// mapping of a reversible circuit using the NC algorithm`
   `   combined with MS algorithm. The obtained circuit contains`
   `   NOT, CNOT, Toffoli, Peres, inverse Peres, and αP gates.`
3 $G' \leftarrow$ ReversibleMapping(G)`//` $G' = T_1(C_1, t_1) \cdots T_k(C_k, t_k)$
4 AddAncilla(a_1, Q)
5 **while** $i < k$ **do** `// Terminate?`
6     $g \leftarrow$ Gate(G', i)
7     $c \leftarrow |C_i|$
8   **if** $c < 2$ **then**
9         $Q \leftarrow Q \circ g$
10   **else if** $c = 2$ **then**
           `// Each Toffoli, Peres, or inverse Peres gate is mapped`
           `   to its equivalent depth optimized quantum circuit`
11   **if** IsToffoli(g) **then**
12            $Q \leftarrow Q \circ$ NewToffoliMapping(g)
13   **else if** IsPeres(g) **then**
14            $Q \leftarrow Q \circ$ NewPeresMapping(g)
15   **else if** IsInversePeres(g) **then**
16            $Q \leftarrow Q \circ$ NewInversePeresMapping(g)
17   **else if** IsAlphaPeres(g) **then**
18            $Q \leftarrow Q \circ$ AlphaPeresMapping(g)
19   **end**
20   **end**
21     $i \leftarrow i + 1$
22 **end**

Fig. 5.12 Application of the local scheme to concurrent Toffoli gates. (**a**) Original circuit. (**b**) Original mapping. (**c**) Proposed mapping

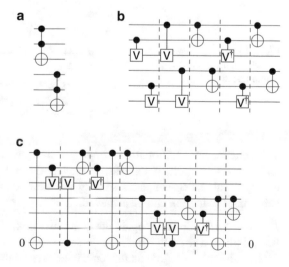

Algorithm 7: Global depth optimization

Input: Quantum circuit $G = T_1(C_1, t_1) \cdots T_k(C_k, t_k)$
Output: Global Depth optimized quantum circuit G
      `// Initialize`
1   $i \leftarrow 0$
2   AddAncilla(a_1, G)
3   **while** $i < k$ **do** `// Terminate?`
        `// find the subcircuit S=`$T_l(c, t_l) \circ \cdots \circ T_k(c, t_k)$` such that all`
        `the gates share the same control line c`
4     $S \leftarrow CommonControl(G, i)$
5   **if** Size(S) > 4 **then**
6       $j \leftarrow 0$
          `// copy the value of the line c to the ancilla `a_1` by`
          `adding the CNOT gate `$T(c, a_1)$` where l is the position`
          `where the gate will be inserted`
7   InsertGate($G, l, T(c, a_1)$)
8   **while** $j <$ Size(S) **do**
9         $T_j(c, t_j) \leftarrow T_j(a_1, t_j)$
10        $j \leftarrow j + 2$
11   **end**
          `// restore the value of the ancilla `a_1` by adding the`
          `CNOT gate `$T(c, a_1)$
12   InsertGate($G, k + 2, T(c, a_1)$)
13   **end**
14     $i \leftarrow k + 3$
15 **end**

Fig. 5.13 Consideration of the whole circuit. (**a**) Original circuit. (**b**) Resulting circuit

Fig. 5.14 Application of global scheme to the circuit from Fig. 5.8a

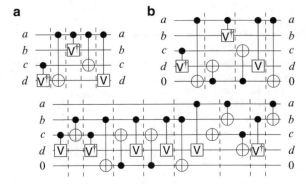

This scheme can additionally be improved by applying the *moving rule* for quantum circuits. As a result, gates can be moved through the circuit which might lead to larger subcircuits of gates sharing the same control line. In this case, a more substantial reduction can be achieved.

Example 5.10. Consider again the quantum circuit shown in Fig. 5.8b. The second, fifth, sixth, and seventh gate share the same control line b *and* can be moved together (note, although the third and the tenth gate have control line b, they cannot be moved to a consecutive circuit). Exploiting that, this circuit can be optimized leading to the circuit shown in Fig. 5.14. This reduces the depth from 12 to 9.

Note that this scheme also increases the NCV-cost of the resulting circuit. However, since for each identified subcircuit only two copy gates have to be added, the increase is almost negligible.

5.2.3   Experimental Results

In order to confirm the efficiency of the proposed idea, the approaches described above have been implemented and experimentally evaluated. For this purpose, the open source toolkit *RevKit* [129] has been applied.

Table 5.1 provides the obtained numbers. For all benchmarks listed in the first column, the number of lines (*Lines*), the NCV-cost (*NCV*), and the depth (*Depth*) of the respective circuit realizations as well as the run-time in seconds (*Time*) required to perform each depth-optimization approach are provided. We distinguish between the following circuits:

- ORIGINAL BENCHMARK (OB) represents the circuits synthesized with the Young subgroups based synthesis approach (YSG [135]), optimized using simulated annealing, and mapped to NCV quantum circuits using the NC mapping algorithm combined with the B2 mapping algorithm as described in Sect. 4.1.1 (see page 92) without any depth-optimization.
- OPTIMIZED CIRCUITS (OC) represent the circuits that have additionally been optimized using the rule-based techniques reviewed in Sect. 5.1.1.1.

Both, the initial circuits and optimized circuits, allow for a comparison to the circuits obtained by the proposed techniques, namely:

- Circuits that have been obtained using the optimization scheme that considers single gates (LOCAL) as described in Sect. 5.2.2.1.
- Circuits that have been obtained using the optimization scheme that considers the whole circuit (GLOBAL) as described in Sect. 5.2.2.2.

The percentage depth improvement of the circuits obtained by the proposed techniques with respect to the original benchmark is provided in the columns denoted by D_{OB}. The D_{LOCAL} denotes the depth-improvement of depth-optimized circuit using local approach with respect to the depth-optimized circuits using the global approach. The percentage NCV-cost decrease after applying our depth-aware approaches with respect to the already NCV-cost optimized circuits is given in the column denoted by NCV_{OC}

Independent of the optimization schemes proposed above, the depth of quantum circuits can be improved using existing optimization schemes that originally aimed for NCV-cost reduction. In particular, the application of reduction rules as already explained in Sect. 5.3 is beneficial. This reduces the NCV-cost but also improves the depth of the circuits with 9 % on average compared to the depth of the original benchmarks. Accordingly, such NCV-cost optimizations are applied to already depth-optimized circuits.

Table 5.1 Experimental results for the depth optimization approaches

	Original benchmark (OB)				Optimized circuits				Local (Sect. 5.2.2.1, +1 ancilla)						Global (Sect. 5.2.2.2, +1 ancilla)						
ID	L	GC	NCV	Depth	NCV	Depth	Time	D_{OB} (%)	L	NCV	Depth	Time	D_{OB} (%)	NCV_{OB} (%)	L	NCV	Depth	Time	D_{OB} (%)	D_{LOCAL} (%)	NCV_{OC} (%)
fredkin	3	3	15	15	13	13	0.00	13.33	4	17	12	0.00	20.00	−23.53	4	15	12	0.00	20.00	0.00	−13.33
3_17	3	6	16	14	14	12	0.00	14.29	4	18	11	0.00	21.43	−22.22	4	14	12	0.00	14.29	−9.09	0.00
ex1	5	20	40	37	20	18	0	51.35	6	20	13	0.01	64.86	0.00	6	22	11	0.00	70.27	15.38	−9.09
sf	4	15	43	37	34	32	0.00	13.51	6	41	25	0.00	32.43	−17.07	6	38	24	0.00	35.14	4.00	−10.53
rd32	4	14	68	63	59	54	0.00	14.29	6	74	44	0.01	30.16	−20.27	6	61	43	0.01	31.75	2.27	−3.28
aj-e11	4	9	87	85	72	70	0.00	17.65	6	88	55	0.01	35.29	−18.18	6	78	58	0.00	31.76	−5.45	−7.69
4gt13	8	29	147	128	123	111	0.01	13.28	10	153	89	0.03	30.47	−19.61	10	125	88	0.03	31.25	1.12	−1.60
4gt11	8	41	148	128	127	110	0.01	14.06	10	169	79	0.04	38.28	−24.85	10	131	84	0.03	34.38	−6.33	−3.05
4mod5	8	35	166	151	156	143	0.01	5.30	10	185	106	0.04	29.80	−15.68	10	166	105	0.04	30.46	0.94	−6.02
alu	8	73	291	252	243	208	0.04	17.46	10	315	145	0.09	42.46	−22.86	10	255	139	0.08	44.84	4.14	−4.71
mini-alu	8	78	291	242	249	208	0.03	14.05	10	311	150	0.10	38.02	−19.94	10	265	142	0.13	41.32	5.33	−6.04
ham7	10	94	308	257	264	219	0.04	14.79	12	320	138	0.21	46.30	−17.50	12	280	126	0.12	50.97	8.70	−5.71
4gt10	8	66	338	299	259	228	0.09	23.75	10	306	171	0.18	42.81	−15.36	10	275	169	0.15	43.48	1.17	−5.82
4gt5	8	81	341	297	298	256	0.06	13.80	10	381	164	0.13	44.78	−21.78	10	314	148	0.14	50.17	9.76	−5.10
4mod7	8	86	367	326	328	295	0.05	9.51	10	377	216	0.16	33.74	−13.00	10	350	206	0.22	36.81	4.63	−6.29
hwb5	8	83	383	331	324	285	0.07	13.90	10	405	212	0.20	35.95	−20.00	10	340	190	0.16	42.60	10.38	−4.71
mod5adder	9	42	400	371	377	352	0.02	5.12	11	417	294	0.29	20.75	−9.59	11	399	277	0.37	25.34	5.78	−5.51

parity	16	240	432	418	240	227	0.28	45.69	17	240	120	0.45	71.29	0.00	17	244	125	0.46	70.10	-4.17	-1.64
decod24-enable	9	77	572	529	524	489	0.07	7.56	11	587	378	0.34	28.54	-10.73	11	542	357	0.51	32.51	5.56	-3.32
cycle10	15	88	577	526	523	492	0.04	6.46	17	562	390	0.40	25.86	-6.94	17	547	375	0.78	28.71	3.85	-4.39
ex3	9	105	815	748	754	699	0.11	6.55	11	838	557	0.97	25.53	-10.02	11	790	536	1.05	28.34	3.77	-4.56
C17	9	73	898	857	856	820	0.06	4.32	11	923	689	0.88	19.60	-7.26	11	884	671	1.22	21.70	2.61	-3.17
cm82a	9	106	899	829	855	791	0.11	4.58	11	932	634	0.95	23.52	-8.26	11	889	590	1.28	28.83	6.94	-3.82
plus63mod8192	16	51	958	919	880	851	0.08	7.40	18	926	746	4.82	18.82	-4.97	18	906	730	6.24	20.57	2.14	-2.87
ex2	9	67	967	922	931	887	0.11	3.80	11	999	735	1.23	20.28	-6.81	11	969	688	1.67	25.38	6.39	-3.92
sym6	10	131	3134	3045	3105	3014	0.11	1.02	12	3178	2618	14.39	14.02	-2.30	12	3193	2475	24.55	18.72	5.46	-2.76
hwb7	10	136	3456	3361	3413	3318	0.18	1.28	12	3509	2823	17.30	16.01	**y-2.74**	12	3501	2722	30.41	19.01	3.58	-2.51
z4	11	290	8885	8661	8851	8630	0.18	0.36	13	8965	7592	129.28	12.34	-1.27	13	9108	7301	238.48	15.70	3.83	-2.82
hwb8	11	298	9615	9385	9599	9369	0.11	0.17	13	9702	8211	160.45	12.51	-1.06	13	9879	7908	285.76	15.74	3.69	-2.83
ham15	18	781	17,639	17,213	17,430	17,027	0.93	1.08	18	17,679	15,310	571.76	11.06	-1.41	18	17,943	14,562	1411.43	15.40	4.89	-2.86
sqn	12	498	20,170	19,770	20,133	19,734	0.14	0.18	14	20,236	17,494	774.22	11.51	-0.51	14	20,579	16,874	1437.81	14.65	3.54	-2.17
sym9	13	1205	60,623	59,505	60,577	59,455	0.27	0.08	13	60,747	52,704	3319.21	11.43	-0.28	13	62,212	50,839	6109.52	14.56	3.54	-2.63
Average								9.53					26.10	-9.74					28.77	3.68	-4.21

It can be observed that our first approach leads to significant improvements (26 % on average, 11 % in the worst case for *ham15*, and up to 71 % in the best case for *parity*). Our second approach enables further improvements. For example, the depth of the benchmark (*hw5*) can be reduced from 331 to 212 (using the local approach from Sect. 5.2.2.1) or 190 (using the global approach from Sect. 5.2.2.2). In the majority of the cases, the global based optimization outperforms the local based optimization with additional reduction of 3.68 and 28 % on average compared to the depth obtained by the local approach and the original benchmark, respectively.

As discussed above, these improvements in the depth may come at the price of higher NCV-cost. As our evaluations show, this particularly holds for the local consideration of single gates (see columns denoted LOCAL). Here, the NCV-cost is increased by 9.36 % on average compared to the already optimized circuit.

Compared to original circuits, improvements of more than 28 % on average, 14 % in the worst case (*3_17*), and 70 % in the best case (*ex1*) can be achieved if the global scheme is applied. These achievements are possible with an increase of 4 % on average in the NCV-cost compared to the optimized circuit. Although this eventually results in the consideration of another qubit to be physically realized, the possible benefits with respect to timing and particularly decoherence time might be worth the overhead.

In summary, by applying local and global methods to the already optimized circuits, significant depth reductions can be achieved keeping less NCV-cost in the meantime according to the original circuits. Further improvement results (3 % on average) can be reached when the global approach is applied comparing to the local approach.

5.3   NCV-Cost Optimization

Optimizing quantum circuits is mainly based on the moving rules applied in a local manner, i.e., quantum gates are rearranged according to specific rules or to line values as reviewed in Sect. 3.1.1.1. To overcome these limitations, an optimization approach based on the equivalence checking aiming to reduce the NCV-cost is proposed. This technique allows to trace the global functionality of the circuit instead of inspecting each line functionality separately from each other. This will guarantee to find all the possible gate rearrangements and thus to apply all possible reductions in the quantum circuit. The application of equivalence checking in the optimization of quantum circuits leads to further NCV-cost reduction. As confirmed by an experimental evaluation, improvements on NCV-cost of 6% on average can be observed compared to the best known NCV-cost optimization approach. The approach has been published in [9].

The remainder of the section is structured as follows. The first section outlines the proposed approach. Then, a detailed description of the implementation of the presented approach is given in Sect. 5.3.2. Finally, experimental results are given in Sect. 5.3.3.

5.3.1 Proposed Idea

Here, we present an approach for reducing quantum gates based on equivalence checking of quantum circuits. To do that, we move the gate in question to the desired position, then we check whether the obtained circuit is implementing the original specification. Thus, we do not have to track each line value, and allow moving the gate only to the position that has the same line value. When the initial circuit is equivalent to the modified one, we apply the deletion or reduction rules. Otherwise, we restore the gate to its original position.

Equivalence checking for quantum functionality has been considered in the past. There are several existing equivalence checking methods for circuits (e.g., based on simulation [143], decision diagrams [87, 124], or Boolean satisfiability [153]). The *QMDD* data structure shows efficiency in terms of memory and time for representing the quantum functionality [99].

We present an approach based on QMDD equivalence checking to optimize quantum circuits. We are introducing an extension of the commonly applied moving rules which allow additional gate rearrangement and thus further gate reductions. As the experiments in Sect. 5.3.3 confirm, the application of our approach leads to further NCV-cost reductions of the resulting quantum circuits.

Figure 5.15 sketches the proposed idea: the gate at position i can be reduced with another gate at position j. To determine the possibility to move the gate at position j after the gate located at position i, we move it to position $i + 1$, then we check if the obtained circuit is equivalent to the original one. The reduction rule is applied to the gates if the circuits are equivalent. Otherwise, the gate j is moved to its original position.

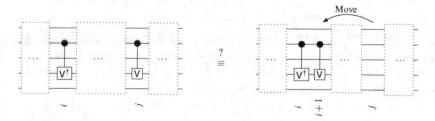

Fig. 5.15 Optimizing quantum circuits based on equivalence checking

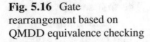 **Fig. 5.16** Gate
rearrangement based on
QMDD equivalence checking

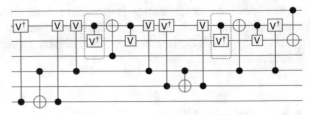

Example 5.11. In the circuit presented in Fig. 5.16, the gates in position 5 and 15 surrounded by red rectangles can cancel each other. We check the possibility of moving the gate on position 5 to position 12 using the QMDD equivalence checking tool. By applying the LLP, the movement is not possible because the target line contains controls and the control line is not invariant between positions 5 and 12. This gate rearrangement cannot also be applied when the DDMF algorithm is adopted because this circuit is containing gates that lead to entangled states.

5.3.2 Application

Equivalence checking of two different quantum circuits is a memory-consuming process for larger circuits. Also, including equivalence checking in the optimization of a circuit is computationally expensive due to the high number of possible gate rearrangements in the circuit. Therefore, we give below three different alternatives to overcome this problem. Then, we show the best scenario with respect to the required runtime in the experimental results.

5.3.2.1 Whole Circuit

In a naive way, the optimization can be applied to the whole quantum circuit by making the equivalence checking of the modified circuit against the original one. To do that, we build the respective QMDD for each circuit, i.e., each one is constructed recursively by multiplying it with each gate QMDD. Since the multiplication is a costly operation, then, the larger the circuit is, the more the time is consumed.

Example 5.12. Figure 5.17 shows a quantum circuit. To determine the possibility of moving the gate in position 15 after the gate located in position 4, we move it to position 5. Then, we construct the QMDD for the original circuit in the green box, as well as the modified one to test their equivalence.

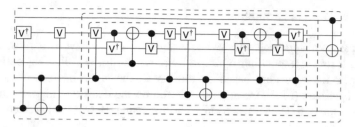

Fig. 5.17 Optimization using different filters

5.3.2.2 Subcircuit

In this case, we take into consideration when building the QMDD only the involved gates instead of considering all the gates of the circuit. In fact, we consider only the subcircuit between the position of the gate that could be moved and the position of the gate that could be reduced with that gate. By adopting this strategy, we decrease the number of multiplications required to build each QMDD.

Example 5.13. Figure 5.17 shows the considered subcircuit in the red box. Only gates between position 4 and 16 are considered.

5.3.2.3 Line Filtered Sub-Circuit

Since the size of the QMDD increases exponentially with the number of lines in a quantum circuit, we consider, in addition to subcircuit restriction, only the used lines in that subcircuit. Hence, we construct the QMDD for the involved gates and lines only.

Example 5.14. Figure 5.17 depicts the line-gate filtered subcircuit in the blue box. Only gates between positions 4 and 16 are considered, and only the lines between 2 and 6 are taken into account since the first and last lines are not used in the subcircuit between positions 4 and 16.

5.3.3 *Experimental Results*

This section provides experimental results for quantum circuit optimization using the QMDD equivalence checking package. The proposed idea has been implemented in the open source toolkit *RevKit* [129] combined with the QMDD package [87]. The experimental evaluation has been carried out using the benchmarks taken from [73], and [147].

In total, two different aspects are studied: the results of our technique (with respect to the NCV-cost) in comparison with previous approaches, and the quality of our approach (with respect to the run-time) when applying different filters described above in Sect. 5.3.2.

5.3.3.1 NCV-Cost Evaluation

To compare our approach with existing algorithms, we have generated for each benchmark its corresponding optimized circuits from the following optimization approaches: the algorithm based on the CMR, the algorithm based on the line labeling approach (LLP), the algorithm based on the decision diagrams for matrix functions approach (DDMF), and our algorithm using the QMDD based equivalence checking (QMDD).

Table 5.2 summarizes the obtained result. All benchmarks are listed in the first column. Then, the number of lines (L), the NCV-cost (NCV_0), the NCV-cost when entanglements are removed (NCV_1). The resulting NCV-cost (NCV) and the needed run-time in seconds (Time) are provided for the different algorithms mentioned above.

The NCV-cost reductions and the relative improvements obtained by the proposed technique compared to the three different approaches (CMR, LLP, and DDMF) are provided in the columns denoted by Δ_{NCV} and I_{NCV}, respectively.

As can be seen, the non-entangled circuits are more expensive compared to the entangled circuits in particular when the original reversible circuit contains MPMCT gates with more than three controls. Only the DDMF approach cannot support the entangled circuit and for that reason, it does not outperform the CRM and the LLP algorithms in several cases because the input circuits to be optimized are initially more expensive.

Over all circuits, our algorithm leads to better reduction in the size of the resulting quantum circuit. The percent improvement of the NCV-cost exceeds 6 % on average compared to the CRM approach, the LLP approach, and the DDMF approach.

5.3.3.2 Time Evaluation

The required run-times for each of the three strategies explained in Sect. 5.3.2 are shown graphically by means of the plot in Fig. 5.18. The values of x-axis and the y-axis (logarithmic scale) denote the benchmark (line, NCV-cost) and the run-time, respectively. The plot contains three different scenarios: the run-time when the whole circuit is considering (dashed-lines bar), the run-time when a subcircuit is considered (white bar), and the run-time when line filtering is considering (dotted bar).

One can clearly see that the best run-time is obtained when the subcircuit alternative is adopted. However considering the whole circuit is a time-consuming task as was expected from the beginning. This returns to the large number of needed multiplication to build a QMDD compared to the subcircuit or line filtered QMDD. In addition, in most of the cases, the line filtered QMDD strategy has a similar run time compared to the subcircuit based one. This scenario does not outperform the subcircuit based technique because of the required time to find the filtered lines each time an equivalence checking is called which can make this strategy be the worst choice when the number of lines in the circuit and the number of built QMDD is large (*alu1*).

Table 5.2 Experimental results for NCV-cost optimization approaches

Original benchmark (OB)				CMR [78]		LLP [86]		DDMF [112]		QMDD		QMDD/CRM		QMDD/LLP		QMDD/DDMF	
ID	L	NCV_0	NCV_1	NCV	Time	NCV	Time	NCV	Time	NCV	Time	Δ_{NCV}	I_{NCV} (%)	Δ_{NCV}	I_{NCV} (%)	Δ_{NCV}	I_{NCV} (%)
alu	5	30	30	30	0.00	30	0.00	30	0.00	28	0.70	2	6.67	2	6.67	2	6.67
rd32	5	124	124	124	0.00	124	0.00	114	0.00	95	1.57	29	23.39	29	23.39	19	16.67
4gt4	6	45	45	45	0.00	45	0.00	44	0.00	43	0.73	2	4.44	2	4.44	1	2.27
sf	6	121	121	121	0.00	121	0.01	108	0.00	97	3.24	24	19.83	24	19.83	11	10.19
majority	7	72	82	72	0.00	72	0.00	80	0.00	70	0.89	2	2.78	2	2.78	10	12.50
ex2	7	100	110	100	0.00	100	0.00	105	0.00	90	1.89	10	10.00	10	10.00	15	14.29
hwb	7	1168	1207	1162	0.02	1160	0.07	1198	0.22	1135	2577.50	27	2.32	25	2.16	63	5.26
cm82a	8	87	87	85	0.00	84	0.00	84	0.00	77	0.82	8	9.41	7	8.33	7	8.33
rd53	8	195	195	189	0.00	188	0.02	186	0.01	178	4.53	11	5.82	10	5.32	8	4.30
rd73	10	421	421	417	0.00	413	0.05	401	0.06	373	16.62	44	10.55	40	9.69	28	6.98
sqn	10	1089	1089	1084	0.01	1084	0.07	1082	0.31	1030	813.98	54	4.98	54	4.98	52	4.81
wim	11	159	159	157	0.00	157	0.01	157	0.01	144	1.91	13	8.28	13	8.28	13	8.28
dc1	11	202	202	196	0.01	196	0.03	196	0.02	185	3.48	11	5.61	11	5.61	11	5.61
z4	11	422	422	422	0.00	419	0.02	416	0.07	395	18966.82	27	6.40	24	5.73	21	5.05
max461	11	1994	2198	1987	0.02	1987	0.15	2182	1.49	1961	92244.68	26	1.31	26	1.31	221	10.13

(continued)

Table 5.2 (continued)

Original benchmark (OB)				CMR [78]		LLP [86]		DDMF [112]		QMDD		QMDD/CRM		QMDD/LLP		QMDD/DDMF	
ID	L	NCV_0	NCV_1	NCV	Time	NCV	Time	NCV	Time	NCV	Time	Δ_{NCV}	I_{NCV} (%)	Δ_{NCV}	I_{NCV} (%)	Δ_{NCV}	I_{NCV} (%)
life	11	2070	2131	2064	0.02	2064	0.11	2089	3.29	1981	68650.71	83	4.02	83	4.02	108	5.17
sqrt8	12	387	387	386	0.00	383	0.04	382	0.06	360	4210.23	26	6.74	23	6.01	22	5.76
cycle10	12	850	914	818	0.03	818	0.16	877	0.25	792	8700.39	26	3.18	26	3.18	85	9.69
radd	13	208	208	208	0.00	208	0.02	204	0.02	197	5.04	11	5.29	11	5.29	7	3.43
plus63mod4096	13	649	703	638	0.02	638	0.07	699	0.11	631	2473.65	7	1.10	7	1.10	68	9.73
pm1	14	233	233	233	0.01	233	0.01	232	0.01	217	12.22	16	6.87	16	6.87	15	6.47
cm85a	14	681	681	679	0.00	677	0.03	674	0.21	642	5154.77	37	5.45	35	5.17	32	4.75
plus63mod8192	14	811	882	798	0.02	798	0.11	864	0.16	789	4923.69	9	1.13	9	1.13	75	8.68
ham15	15	198	198	198	0.01	198	0.01	198	0.01	189	2.90	9	4.55	9	4.55	9	4.55
ham15	15	358	358	358	0.00	356	0.01	355	0.04	346	35.71	12	3.35	10	2.81	9	2.54
ham15	15	1161	1161	1155	0.01	1155	0.08	1134	0.51	1122	6909.74	33	2.86	33	2.86	12	1.06
co14	15	1290	1458	1290	0.00	1290	0.04	1458	0.45	1290	17129.88	0	0.00	0	0.00	168	11.52
sqr6	18	804	804	801	0.01	800	0.06	783	0.49	763	16.00	38	4.74	37	4.63	20	2.55
alu1	20	189	189	189	0.01	189	0.01	184	0.02	173	8.52	16	8.47	16	8.47	11	5.98
pcler8	21	291	291	291	0.01	291	0.01	279	0.04	258	17.43	33	11.34	33	11.34	21	7.53
mux	22	541	541	535	0.02	533	0.04	532	0.14	514	11568.29	21	3.93	19	3.56	18	3.38
Average												22	6.28	21	6.11	37	6.91

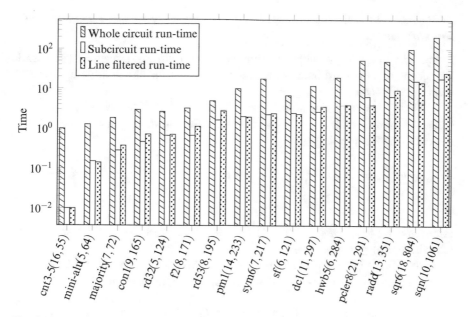

Fig. 5.18 Time evaluation of the used filters

5.4 Complexity Analysis of Quantum Circuits

We provide an extensive overview of upper bounds on the number of gates needed in quantum implementation of reversible gates as well as reversible circuits. The study has been published in [7].

5.4.1 Complexity of NCV Quantum Circuits

This section first gives upper bounds for the realization of MPMCT and ST gates in terms of NCV gates and then gives upper bounds for reversible circuits based on NCV gates.

5.4.1.1 NCV-Cost for MPMCT Gates

Theorem 5.1. *A c-control MPMCT gate with $c \geq 5$ can be realized with at most*

- $um_{\text{NCV}_{\text{B1}}}(c) \leq 24(c-3) + 12$
- $um_{\text{NCV}_{\text{NC}}}(c) \leq 18(c-3) + 10$
- $um_{\text{NCV}_{\text{MI}}}(c) \leq 24(c-4) + 16$
- $um_{\text{NCV}_{\text{B2}}}(c) \leq 12(c-2) + 2$

NCV gates. When the MPMCT gate has only negative controls four additional NOT gates are needed.

Proof. The first and last upper bounds have been already proven in [83], while the second and the third bounds are calculated using the circuit structure for each mapping approach.

In the NC mapping, two of the three resulting MPMCT gates have $\lceil \frac{c}{2} \rceil$ controls while the third one has $c + 1 - \lceil \frac{c}{2} \rceil$ controls. Each of these gates is mapped with respect to the B2 mapping, giving

$$um_{NCV_{NC}}(c) \leq 2\left(12\left(\left\lceil \tfrac{c}{2}\right\rceil - 2\right) + 2\right) + 12\left(c + 1 - \left\lceil \tfrac{c}{2}\right\rceil - 2\right) + 2 + 4$$

$$\leq 12\left(c + \left\lceil \tfrac{c}{2}\right\rceil - 5\right) + 10$$

For even c, we have $um_{NCT_{NC}}(c) \leq 18\left(c - \frac{10}{3}\right) + 10 \leq 18(c - 3) + 10$, and for odd c, we have $u(c) \leq 18(c - 3) + 10$.

In the MI mapping two of the resulting four MPCMCT gates have $\lceil \frac{c}{2} \rceil$ controls while the other two have $c - \lceil \frac{c}{2} \rceil$ controls. Applying B2 mapping yields

$$um_{NCT_{MI}}(c) \leq 2\left(12\left(\left\lceil \tfrac{c}{2}\right\rceil - 2\right) + 2\right) + 2\left(12\left(c - \left\lceil \tfrac{c}{2}\right\rceil - 2\right) + 2\right) + 4 + 4$$

$$\leq 24(c - 4) - 16$$

Toffoli gates.   ∎

Table 5.3 summarizes the above theorem, along with the number of ancillae used by each mapping approach.

5.4.1.2   NCV-Cost for Single-Target Gates

Corollary 5.1. *From Lemma 4.4, we derive the following inequations for $n \geq 6$:*

$$\sum_{i=0}^{\left\lceil \frac{n}{2}\right\rceil}(12(i - 2) + 2) \cdot \binom{n}{i} + \sum_{i=\left\lceil \frac{n}{2}\right\rceil + 1}^{n}(24(i - 3) + 12) \cdot \binom{n}{i} \leq (18n - 63) \cdot 2^{n-1}$$

(5.3)

Table 5.3 NCV-cost for an MPMCT gate with c controls

Mapping	Ancillae	NCV
B1 (Barenco et al. [15, Lemma 7.3])	1	$24(c - 3) + 12$ [83]
NC (Nielsen and Chuang [98])	1[a]	$18(c - 3) + 10$
MI (Miller et al. [91])	1	$24(c - 4) + 16$
B2 (Barenco et al. [15, Lemma 7.2])	$c - 2$	$12(c - 2) + 2$ [83]

[a]The ancilla is initialized to 0

$$\sum_{i=0}^{\left\lceil \frac{n}{2} \right\rceil}(12(i-2)+2)\cdot\binom{n}{i} + \sum_{i=\left\lceil \frac{n}{2} \right\rceil+1}^{n}(18(i-3)+10)\cdot\binom{n}{i} \le (15n-55)\cdot 2^{n-1}$$

$$(5.4)$$

$$\sum_{i=0}^{\left\lceil \frac{n}{2} \right\rceil}(12(i-2)+2)\cdot\binom{n}{i} + \sum_{i=\left\lceil \frac{n}{2} \right\rceil+1}^{n}(24(i-4)+16)\cdot\binom{n}{i} \le (18n-72)\cdot 2^{n-1}$$

$$(5.5)$$

Proof. We consider the cases when n is even and n is odd. n **is even:** we have

$$\left\lceil \frac{n}{2} \right\rceil = \frac{n}{2}$$

$$\sum_{i=0}^{\frac{n}{2}}\binom{n}{i} = 2^{n-1} + \frac{1}{2}\binom{n}{\frac{n}{2}}$$

$$\sum_{i=\frac{n}{2}+1}^{n}\binom{n}{i} = 2^{n-1} - \frac{1}{2}\binom{n}{\frac{n}{2}}$$

Then,

$$\sum_{i=0}^{\frac{n}{2}}(12(i-2)+2)\cdot\binom{n}{i} + \sum_{i=\frac{n}{2}+1}^{n}(24(i-3)+12)\cdot\binom{n}{i}$$

$$= 12\cdot\sum_{i=0}^{\frac{n}{2}}i\binom{n}{i} - 22\cdot\sum_{i=0}^{\frac{n}{2}}\binom{n}{i} + 24\cdot\sum_{i=\frac{n}{2}+1}^{n}i\binom{n}{i} - 60\cdot\sum_{i=\frac{n}{2}+1}^{n}\binom{n}{i}$$

$$\overset{\text{Lemma 4.4}}{=} 12n\cdot 2^{n-2} - 22\left(2^{n-1} + \frac{1}{2}\binom{n}{\frac{n}{2}}\right)$$

$$+ 24n\cdot 2^{n-2} - 60\left(2^{n-1} - \frac{1}{2}\binom{n}{\frac{n}{2}}\right)$$

$$= (18n-82)\cdot 2^{n-1} + 19\binom{n}{\frac{n}{2}}$$

We have

$$\binom{n}{\frac{n}{2}} \leq 2^{n-1}$$

Hence,

$$\sum_{i=0}^{\frac{n}{2}} (12(i-2)+2) \cdot \binom{n}{i} + \sum_{i=\frac{n}{2}+1}^{n} (24(i-3)+12) \cdot \binom{n}{i}$$

$$\leq (18n-82) \cdot 2^{n-1} + 19.2^{n-1} \leq (18n-63) \cdot 2^{n-1}$$

n is odd: we have

$$\left\lceil \frac{n}{2} \right\rceil = \frac{n+1}{2}$$

$$\sum_{i=0}^{\frac{n+1}{2}} \binom{n}{i} = 2^{n-1} + \binom{n}{\frac{n+1}{2}}$$

$$\sum_{i=\frac{n+3}{2}}^{n} \binom{n}{i} = 2^{n-1} - \binom{n}{\frac{n+1}{2}}$$

Then

$$\sum_{i=0}^{\frac{n+1}{2}} (12(i-2)+2) \cdot \binom{n}{i} + \sum_{i=\frac{n+3}{2}}^{n} (24(i-3)+12) \cdot \binom{n}{i}$$

$$= 12 \cdot \sum_{i=0}^{\frac{n+1}{2}} i \binom{n}{i} - 22 \cdot \sum_{i=0}^{\frac{n+1}{2}} \binom{n}{i} + 24 \cdot \sum_{i=\frac{n+3}{2}}^{n} i \binom{n}{i} - 60 \cdot \sum_{i=\frac{n+3}{2}}^{n} \binom{n}{i}$$

$$\overset{\text{Lemma 4.4}}{=} 12 \left(n \cdot 2^{n-2} + \frac{n+1}{4} \binom{n}{\frac{n+1}{2}} \right) - 22 \left(2^{n-1} + \binom{n}{\frac{n+1}{2}} \right)$$

$$+ 24 \left(n \cdot 2^{n-2} - \frac{n+1}{4} \binom{n}{\frac{n+1}{2}} \right) - 60 \left(2^{n-1} - \binom{n}{\frac{n+1}{2}} \right)$$

$$= (18n-82) \cdot 2^{n-1} - (3n-35) \binom{n}{\frac{n+1}{2}}$$

$$\leq (18n-63) \cdot 2^{n-1} - 19 \cdot 2^{n-1} - (3n-35) \binom{n}{\frac{n+1}{2}}$$

Since $\binom{\frac{n+1}{2}}{n} \leq 3 \cdot 2^{n-3}$, we get

$$\sum_{i=1}^{\frac{n+1}{2}}(12(i-2)+2) \cdot \binom{n}{i} + \sum_{i=\frac{n+3}{2}}^{n}(24(i-3)+12) \cdot \binom{n}{i}$$

$$\leq (18n-63) \cdot 2^{n-1} - 19 \cdot 2^{n-1} - (3n-35) \cdot 3 \cdot 2^{n-3}$$

$$= (18n-63) \cdot 2^{n-1} - (9n-29) \cdot 2^{n-3} \leq (18n-63) \cdot 2^{n-1}$$

The other inequations are obtained using the same argument.   ∎

Theorem 5.2. *An n-variable single-target gate can be realized, if $n \geq 6$, with at most*

- $us_{\mathrm{NCV_{B1}}}(n) \leq (18n-81) \cdot 2^{n-2}$
- $us_{\mathrm{NCV_{NC}}}(n) \leq (15n-70) \cdot 2^{n-2}$
- $us_{\mathrm{NCV_{MI}}}(n) \leq (18n-90) \cdot 2^{n-2}$

NCV gates.

Proof. Using the B1 mapping, we have

$$us_{\mathrm{NCV_{B1}}}(c) \leq \sum_{i=0}^{\lceil\frac{n-1}{2}\rceil}(12(i-2)+2)\binom{n-1}{i} + \sum_{i=\lceil\frac{n-1}{2}\rceil+1}^{n-1}(24(i-3)+12)\binom{n-1}{i}$$

$$\overset{\text{Corollary 5.1}}{\leq} (18(n-1)-63) \cdot 2^{n-2}$$

$$= (18n-81) \cdot 2^{n-2}$$

Using the NC mapping, we have

$$us_{\mathrm{NCV_{NC}}}(c) \leq \sum_{i=0}^{\lceil\frac{n-1}{2}\rceil}(12(i-2)+2)\binom{n-1}{i} + \sum_{i=\lceil\frac{n-1}{2}\rceil+1}^{n-1}(18(i-3)+10)\binom{n-1}{i}$$

$$\overset{\text{Corollary 5.1}}{\leq} (15(n-1)-55) \cdot 2^{n-2}$$

$$= (15n-70) \cdot 2^{n-2}$$

Using the MI mapping, we have

$$us_{NCV_{MI}}(c) \leq \sum_{i=0}^{\lceil \frac{n-1}{2} \rceil} (12(i-2) + 2) \binom{n-1}{i} + \sum_{i=\lceil \frac{n-1}{2} \rceil + 1}^{n-1} (24(i-4) + 16) \binom{n-1}{i}$$

$$\overset{\text{Corollary 5.1}}{\leq} (18(n-1) - 72) \cdot 2^{n-2}$$

$$= (18n - 90) \cdot 2^{n-2}$$

∎

5.4.1.3 NCV-Cost for Quantum Circuits

In the following, we present the complexity of NCV circuits using the transformation based synthesis approach with different mapping strategies.

Theorem 5.3. *Using the transformation based synthesis algorithm, an n qubit NCV circuit with $n \geq 6$ has the following upper bound on the number of NCV gates:*

- $uc_{NCV_{B1}}(n) \leq (11n - 33) \cdot 2^n$
- $uc_{NCV_{NC}}(n) \leq (9.5n - 29.5) \cdot 2^n$
- $uc_{NCV_{MI}}(n) \leq (11n - 38) \cdot 2^n$

Proof. A reversible circuit synthesized with the transformation based approach will have at most n NOT gates, $(n-1) \cdot 2^{n+1} - n^2 + 4$ CNOT gates, and $\sum_{i=2}^{n-1} \binom{n}{i}$ MCT gates where i denotes the number of controls on each gate.

Based on the B1 mapping, we see that

$$uc_{NCT_{B1}}(n) \leq n + (n-1) \cdot 2^{n+1} - n^2 + 4 + \sum_{i=2}^{\lceil \frac{n}{2} \rceil} (12(i-2) + 2) \cdot \binom{n}{i}$$

$$+ \sum_{i=\lceil \frac{n}{2} \rceil + 1}^{n-1} (24(i-3) + 12) \cdot \binom{n}{i}$$

$$\overset{\text{Corollary 5.1}}{\leq} n + (n-1) \cdot 2^{n+1} - n^2 + 4 + (18n - 63) \cdot 2^{n-1}$$

$$\leq (11n - 33) \cdot 2^n + n - n^2 + 4$$

$$\leq (11n - 33) \cdot 2^n$$

The other upper bounds are calculated using the same method.   ∎

Table 5.4 Updated T-depth for an MPMCT gate with c controls

Mapping	Ancillae	Clifford $+ T$
B1 (Barenco et al. [15, Lemma 7.3])	1	$8(c-2)$
NC (Nielsen and Chuang [98])	1^a	$6(c-2)+2$
MI (Miller et al. [91])	1	$8(c-3)+4$
B2 (Barenco et al. [15, Lemma 7.2])	$c-2$	$4(c-1)$

[a] The ancilla is initialized to 0

5.4.2 Complexity of Clifford+T Quantum Circuits

This section first presents the existing upper bounds on the T-depth of MPMCT gates. Afterwards, new upper bounds on the T-depth required to realize MPMCT gate, ST gate, and quantum circuits are then given.

5.4.2.1 T-Depth for MPMCT Gates

Table 5.4 summarizes the upper bounds for the T-depth of each mapping defined in Sect. 4.1.1 when applied to an MPMCT gate with c control lines for $c \geq 5$. These upper bounds are obtained by directly mapping each two-control Toffoli to an (optimal) T-depth 3 circuit. Likewise, we map the controlled V to an optimal T-depth 2 circuit. In the following, we derive better upper bounds by making use of T gate cancellations.

Theorem 5.4. *A c-control MPMCT gate with $c \geq 5$ controls can be realized with T-depth at most*

- $um_{\text{CLF}_{B1}}(c) \leq 8(c-2)$
- $um_{\text{CLF}_{NC}}(c) \leq 6(c-2)+2$
- $um_{\text{CLF}_{MI}}(c) \leq 8(c-3)+4$
- $um_{\text{CLF}_{B2}}(c) \leq 4(c-1)$

Proof. Please refer to the proofs of Lemmas 4.2 and 4.3 ∎

5.4.2.2 T-Depth for Single-Target Gates

Corollary 5.2. *From Lemma 4.4, we derive the following inequations for $n \geq 6$:*

$$\sum_{i=0}^{\lceil \frac{n}{2} \rceil} 4(i-1) \cdot \binom{n}{i} + \sum_{i=\lceil \frac{n}{2} \rceil+1}^{n} 8(i-2) \cdot \binom{n}{i} \leq (6n-14) \cdot 2^{n-1} \quad (5.6)$$

$$\sum_{i=0}^{\left\lceil \frac{n}{2} \right\rceil} 4(i-1) \cdot \binom{n}{i} + \sum_{i=\left\lceil \frac{n}{2} \right\rceil +1}^{n} (6(i-2)+2) \cdot \binom{n}{i} \leq (5n-11) \cdot 2^{n-1} \qquad (5.7)$$

$$\sum_{i=0}^{\left\lceil \frac{n}{2} \right\rceil} 4(i-1) \cdot \binom{n}{i} + \sum_{i=\left\lceil \frac{n}{2} \right\rceil +1}^{n} (8(i-3)+4) \cdot \binom{n}{i} \leq (6n-16) \cdot 2^{n-1} \qquad (5.8)$$

Proof. We consider the cases when n is even and n is odd. n **is even:** we have

$$\left\lceil \frac{n}{2} \right\rceil = \frac{n}{2}$$

$$\sum_{i=0}^{\frac{n}{2}} \binom{n}{i} = 2^{n-1} + \frac{1}{2}\binom{n}{\frac{n}{2}}$$

$$\sum_{i=\frac{n}{2}+1}^{n} \binom{n}{i} = 2^{n-1} - \frac{1}{2}\binom{n}{\frac{n}{2}}$$

Then,

$$\sum_{i=0}^{\frac{n}{2}} 4(i-1) \cdot \binom{n}{i} + \sum_{i=\frac{n}{2}+1}^{n} 8(i-2) \cdot \binom{n}{i}$$

$$= 4 \cdot \sum_{i=0}^{\frac{n}{2}} i\binom{n}{i} - 4 \cdot \sum_{i=0}^{\frac{n}{2}} \binom{n}{i} + 8 \cdot \sum_{i=\frac{n}{2}+1}^{n} i\binom{n}{i} - 16 \cdot \sum_{i=\frac{n}{2}+1}^{n} \binom{n}{i}$$

$$\overset{\text{Lemma 4.4}}{=} 4n \cdot 2^{n-2} - 4\left(2^{n-1} + \frac{1}{2}\binom{n}{\frac{n}{2}}\right) + 8n \cdot 2^{n-2} - 16\left(2^{n-1} - \frac{1}{2}\binom{n}{\frac{n}{2}}\right)$$

$$= (6n-20) \cdot 2^{n-1} + 6\binom{n}{\frac{n}{2}}$$

We have

$$\binom{n}{\frac{n}{2}} \leq 2^{n-1}$$

Hence,

$$4 \cdot \sum_{i=0}^{\frac{n}{2}} (i-2) \binom{n}{i} + 8 \cdot \sum_{i=\frac{n}{2}+1}^{n} (i-3) \binom{n}{i}$$

$$\leq (6n - 20) \cdot 2^{n-1} + 6 \cdot 2^{n-1}$$

$$\leq (6n - 14) \cdot 2^{n-1}$$

n is odd: we have

$$\left\lceil \frac{n}{2} \right\rceil = \frac{n+1}{2}$$

$$\sum_{i=0}^{\frac{n+1}{2}} \binom{n}{i} = 2^{n-1} + \binom{n}{\frac{n+1}{2}}$$

$$\sum_{i=\frac{n+3}{2}}^{n} \binom{n}{i} = 2^{n-1} - \binom{n}{\frac{n+1}{2}}$$

Then,

$$\sum_{i=0}^{\frac{n+1}{2}} 4(i-2) \cdot \binom{n}{i} + \sum_{i=\frac{n+3}{2}}^{n} 8(i-2) \cdot \binom{n}{i}$$

$$= 4 \cdot \sum_{i=0}^{\frac{n+1}{2}} i \binom{n}{i} - 8 \cdot \sum_{i=0}^{\frac{n+1}{2}} \binom{n}{i} + 8 \cdot \sum_{i=\frac{n+3}{2}}^{n} i \binom{n}{i} - 16 \cdot \sum_{i=\frac{n+3}{2}}^{n} \binom{n}{i}$$

Using Lemma 4.4, we obtain

$$\sum_{i=0}^{\frac{n+1}{2}} 4(i-2) \cdot \binom{n}{i} + \sum_{i=\frac{n+3}{2}}^{n} 8(i-2) \cdot \binom{n}{i}$$

$$= 4 \left(n \cdot 2^{n-2} + \frac{n+1}{4} \binom{n}{\frac{n+1}{2}} \right) - 8 \left(2^{n-1} + \binom{n}{\frac{n+1}{2}} \right)$$

$$+ 8 \left(n \cdot 2^{n-2} - \frac{n+1}{4} \binom{n}{\frac{n+1}{2}} \right) - 16 \left(2^{n-1} - \binom{n}{\frac{n+1}{2}} \right)$$

$$= (6n - 20) \cdot 2^{n-1} - (2n - 18) \binom{n}{\frac{n+1}{2}}$$

$$\leq (6n - 20) \cdot 2^{n-1} - (2n - 18) \cdot 2^{n-1}$$

$$\leq (6n - 14) \cdot 2^{n-1} - (2n - 12) \cdot 2^{n-1} \leq (6n - 14) \cdot 2^{n-1}$$

The other inequations are obtained using the same argument. ∎

Theorem 5.5. *An n-variable single-target gate with $n \geq 6$ can be realized with a T-depth of at most*

- $us_{\mathrm{CLF_{B1}}}(n) \leq (6n - 20) \cdot 2^{n-2}$
- $us_{\mathrm{CLF_{NC}}}(n) \leq (5n - 16) \cdot 2^{n-2}$
- $us_{\mathrm{CLF_{MI}}}(n) \leq (6n - 22) \cdot 2^{n-2}$

Proof. Based on the B1 mapping, we obtain

$$us_{\mathrm{CLF_{B1}}}(n) \leq \sum_{i=0}^{\lceil \frac{n-1}{2} \rceil} 4(i-1) \cdot \binom{n-1}{i} + \sum_{i=\lceil \frac{n-1}{2} \rceil + 1}^{n-1} 8(i-2) \cdot \binom{n-1}{i}$$

$$\overset{\text{Corollary 5.2}}{\leq} (6(n-1) - 14) \cdot 2^{n-2}$$

$$\leq (6n - 20) \cdot 2^{n-2}$$

Based on the NC mapping, we obtain

$$us_{\mathrm{CLF_{NC}}}(n) \leq \sum_{i=0}^{\lceil \frac{n-1}{2} \rceil} 4(i-1) \cdot \binom{n-1}{i} + \sum_{i=\lceil \frac{n-1}{2} \rceil + 1}^{n-1} 6(i-2) \cdot \binom{n-1}{i}$$

$$\overset{\text{Corollary 5.2}}{\leq} (5(n-1) - 11) \cdot 2^{n-2}$$

$$\leq (5n - 16) \cdot 2^{n-2}$$

Based on the MI mapping, we obtain

$$us_{\mathrm{CLF_{MI}}}(n) \leq \sum_{i=0}^{\lceil \frac{n-1}{2} \rceil} 4(i-1) \cdot \binom{n-1}{i} + \sum_{i=\lceil \frac{n-1}{2} \rceil + 1}^{n-1} (8(i-3) + 4) \cdot \binom{n-1}{i}$$

$$\overset{\text{Corollary 5.2}}{\leq} (6(n-1) - 16) \cdot 2^{n-2}$$

$$\leq (6n - 22) \cdot 2^{n-2}$$

∎

5.4.2.3   *T*-Depth for Quantum Circuits

So far, the studies of the upper bounds on reversible quantum circuits have been given on circuits over the MCT, NCT, and NCV libraries, but not over the Clifford + *T* library. In this section we study and compare the *T*-depth of such circuits.

Theorem 5.6. *Using the transformation based synthesis approach, an n-variable reversible function can be realized over* Clifford + *T with a T-depth at most*

- $uc_{\text{CLF}_{\text{B1}}}(n) \leq (6n - 14) \cdot 2^{n-1}$
- $uc_{\text{CLF}_{\text{NC}}}(n) \leq (5n - 11) \cdot 2^{n-1}$
- $uc_{\text{CLF}_{\text{MI}}}(n) \leq (6n - 16) \cdot 2^{n-1}$

Proof. For an *n*-variable reversible function, the synthesis approach produces a circuit with at most n NOT gates, $((n-1)\cdot 2^{n+1} - n^2 + 4)$ CNOT gates, and $\sum_{i=2}^{n-1} \binom{n}{i}$ MCT gates where i denotes the number of controls on each gate. The *T*-depth of a NOT or a CNOT gate is 0.

Based on the B1 mapping, we obtain at most

$$uc_{\text{CLF}_{\text{B1}}}(n) \leq \sum_{i=2}^{\left\lceil \frac{n}{2} \right\rceil} 4(i-1) \cdot \binom{n}{i} + \sum_{i=\left\lceil \frac{n}{2} \right\rceil + 1}^{n-2} 8(i-2) \cdot \binom{n}{i}$$

$$\overset{\text{Corollary 5.2}}{\leq} \quad (6n - 14) \cdot 2^{n-1}$$

Based on the NC mapping, we obtain at most

$$uc_{\text{CLF}_{\text{NC}}}(n) \leq \sum_{i=2}^{\left\lceil \frac{n}{2} \right\rceil} 4(i-1) \cdot \binom{n}{i} + \sum_{i=\left\lceil \frac{n}{2} \right\rceil + 1}^{n-2} 6(i-2) \cdot \binom{n}{i}$$

$$\overset{\text{Corollary 5.2}}{\leq} \quad (5n - 7) \cdot 2^{n-1}$$

Based on the MI mapping, we obtain at most

$$uc_{\text{CLF}_{\text{MI}}}(n) \leq \sum_{i=2}^{\left\lceil \frac{n}{2} \right\rceil} 4(i-1) \cdot \binom{n}{i} + \sum_{i=\left\lceil \frac{n}{2} \right\rceil + 1}^{n-2} (8(i-3) + 4) \cdot \binom{n}{i}$$

$$\overset{\text{Corollary 5.2}}{\leq} \quad (6n - 14) \cdot 2^{n-1}$$

∎

5.5   Summary

In this section, we introduced depth optimization for NCV quantum circuits based on an additional ancilla. Two approaches, namely gate based and circuit based, have been considered. Experimental results for the two methods have shown significant depth reductions which reach over 28 % for quantum circuits. Although these methods increase the NCV-cost, applying further improvements to the quantum circuits has fixed the problem.

Afterward, we proposed an optimization approach of quantum circuits based on the equivalence checking. We observed that the existing moving rules do not lead to find all the possible reductions in the circuit. Motivated by this, we introduced an improved optimization approach which exploits the equivalence checking when finding gates to be reduced. Improvements of 6 % on average in the NCV-cost have been observed compared to the LLP and the DDMF based optimization approaches.

Finally, we studied the complexity of the quantum implementations of reversible circuits and gave upper bounds for MPMCT gates, ST gates, and quantum circuits that consist of the NCV or the Clifford $+ T$ gates.

Chapter 6
Conclusions

In this book several approaches have been presented to optimize quantum circuits as well as to analyze their complexity on the three major level of the design flow depicted in Fig. 6.1. The achieved contributions on each level have been described in separate chapters. Accordingly, on each level of the design flow, two approaches have been described to improve the cost of the final realization. Besides, an extensive study of the complexity of implementing a given function has been outlined.

In Chap. 3, a post-synthesis optimization technique has been proposed that reduces the quantum cost of the circuit by applying an efficient template matching algorithm. The question "find all subcircuits that can be replaced by cheaper alternatives" has been formulated as a satisfiability instance, therefore ensuring to finding all possible reductions. For this purpose, SMT solvers have been exploited. Experiments show that despite the exponential complexity of the addressed problem, the given technique is effective particularly for small circuits. By applying the proposed approach, the circuit cost can be reduced by 19 % on average. For large circuits, simulated annealing has been applied to reduce the quantum cost. The given algorithm picks up random gates from a circuit and checks for possible reductions using the rewriting rules that allow more freedom in rearranging gates inside a circuit. The application of this strategy compared to existing greedy algorithms has shown its efficiency to generate optimized results with an improvement of 17 % on average compared to the currently known approaches. Later on, the complexity of circuits in the reversible level has been considered. Better upper bounds for MCT as well as MPMCT based reversible circuits have been determined. The given bounds have been derived based on the Young subgroups synthesis approach.

In Chap. 4, the mapping of reversible circuits into NCT circuits has been illustrated by two approaches. The first approach proposes an improved mapping algorithm of reversible circuits that consist of ST gates. Since each ST gate contains a Boolean control function, the given method attempts to find a decomposition based on its BDD representation. It consists on breaking large ST gates into smaller ones using three additional lines. This enables significant reductions in the quantum cost

© Springer International Publishing Switzerland 2016
N. Abdessaied, R. Drechsler, *Reversible and Quantum Circuits*,
DOI 10.1007/978-3-319-31937-7_6

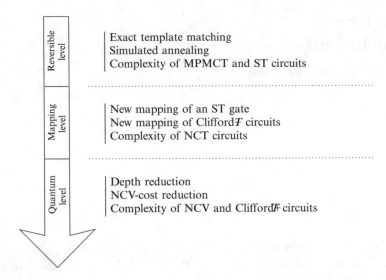

Fig. 6.1 Book contributions

of resulting circuits. Experiments show that considering Boolean decomposition in the mapping of ST gates leads to improvements of 20 % on average in the resulting quantum cost compared to standard mapping algorithm. The second approach presents an improved mapping of MPMCT circuits into Clifford + T quantum circuits. Here, the existing mapping algorithms (B1, B2, NC, and MI) have been optimized in order to translate MPMCT gates into reduced T-depth Clifford + T circuits. Experiments show that the outlined approach leads to a significant T-depth reduction of 50 % on average compared to existing mapping algorithms. Next, upper bounds on the number of NCT gates needed to realize an MPMCT gate, an ST gate, and a general NCT reversible circuit have been given.

In Chap. 5, an optimization approach that aims for the reduction of depth in the process of mapping a reversible circuit into a quantum circuit has been introduced. For this purpose, an improved (local) mapping of single gates as well as a (global) optimization scheme considering the whole circuit have been presented. In both cases, we incorporate the idea of exploiting additional circuit lines which have been used in order to split a chain of serial gates. Our optimization techniques enable a concurrent application of gates which significantly reduces the depth of the circuit. Experiments show that reductions of approximately 28 % on average can be achieved when following this scheme. Then, a second optimization approach has been described in which an NCV circuit resulting from a technology mapping approach has been optimized in terms of the quantum cost. Here, the use of the QMDD based equivalence checking is more efficient in optimizing NCV circuits than the existing optimization approaches. Experiments illustrate that the proposed algorithm indeed is able to efficiently generate cheaper NCV-cost with an improvements of 6 % on average compared to the LLP and the DDMF based

Table 6.1 Complexity of reversible and quantum circuits

MPMCT complexity Sect. 3.4.4	Technology mapping Sect. 4.1.1	NCT complexity Sect. 4.4.3	NCV complexity Sect. 5.4.1.3	Clifford + T complexity Sect. 5.4.2.3
$3(2n - 1) \cdot 2^{n-4}$ [4]	B1 [15]	$(10n - 28) \cdot 2^{n-1}$	$(11n - 33) \cdot 2^n$	$(6n - 14) \cdot 2^{n-1}$
	NC^a [98]	$(9n - 25) \cdot 2^{n-1}$	$(9.5n - 29.5) \cdot 2^n$	$(5n - 11) \cdot 2^{n-1}$
	MI [91]	$(10n - 32) \cdot 2^{n-1}$	$(11n - 38)n \cdot 2^n$	$(6n - 16) \cdot 2^{n-1}$

[a] The used ancilla is initialized to 0

Table 6.2 Summary for gate and circuit complexity

Gate library	MPMCT gate	ST gate	Circuit
ST	0	1	$2n - 1$
MPMCT	1	$3 \cdot 2^{n-4}$	$3(2n - 1) \cdot 2^{n-4}$
NCT	$6(c - 3)$	$(5n - 26) \cdot 2^{n-2}$	$(9n - 25) \cdot 2^{n-1}$
NCV	$18(c - 3) + 6$	$(15n - 70) \cdot 2^{n-2}$	$8.5n \cdot 2^n$ [108]
Clifford + T	$6(c - 2) + 2$	$(5n - 16) \cdot 2^{n-2}$	$(5n - 11) \cdot 2^{n-1}$

optimization approaches. Finally, upper bounds for MPMCT gates, ST gates, and quantum circuits that consist of the NCV or the Clifford + T gates have been presented. Table 6.1 outlines the upper bounds for reversible and quantum circuits using the mapping methods B1, NC, and MI. Note that the table sketches the circuit upper bounds derived from the transformation based synthesis approach [88]. Table 6.2 summarizes the best known upper bounds to represent an MPMCT gate (first column), a single-target gate (second column), and a circuit (third column) using ST gates, MPMCT gates, NCT gates, NCV gates, and Clifford + T gates.

As can be seen from Tables 6.1 and 6.2, a comprehensive overview of upper bounds for the most commonly used reversible and quantum gate libraries has been proposed in this book. Most of the bounds are tighter than the previously known ones. The bounds can be used to evaluate the quality of synthesis approaches. Although several bounds have been proposed, concrete circuit realizations for worst case functions are still missing. Knowing such functions would be very helpful for evaluating the limitations of known and new synthesis approaches.

In contrast to electronic circuits, there exist no standard realization of a quantum computer; for two different realizations of quantum computers, different libraries of gates are applicable with different cost metrics. Each library requires additional constraints in the circuit such as having the nearest-neighbor on NCV circuits and reduced T-depth on Clifford+T circuits. Incorporating these constraints in quantum circuits need to be considered. Our future work in the field of reversible and quantum circuits optimization will focus first on determining efficient algorithms to reduce the T-depth of quantum circuit in the quantum level, and second to find cost-aware methods to transform quantum circuits from a given quantum gate library to another. We will investigate the different methodologies to obtain cheaper Clifford + T circuits. Since our QMDD based approach guarantee an optimal NCV-cost reduction for NCV circuits, we will study the possibility of mapping NCV circuits into Clifford + T circuits and compare it with the best existing mapping algorithm for mapping reversible circuits into Clifford+ T circuit. Then, we will aim at finding an improved algorithm to reduce the quantum cost of Clifford + T circuits as well as their depth.

As a next step in the complexity analysis of reversible and quantum circuits, we will focus on the classification of reversible functions and investigate its application to circuit complexity. In particularly, we will consider the effect of negating and permuting input and output variables. Then, we will compute the number of resulting equivalence classes and study the complexity of reversible circuits for functions in the same class. We also will extend our considerations for linear and affine transformations.

References

1. Abdessaied, N., Soeken, M., Wille, R., Drechsler, R.: Exact template matching using Boolean satisfiability. In: International Symposium on Multiple-Valued Logic, pp. 328–333. IEEE, New York (2013)
2. Abdessaied, N., Wille, R., Soeken, M., Drechsler, R.: Reducing the depth of quantum circuits using additional circuit lines. In: Reversible Computation, pp. 221–233. Springer, New York (2013)
3. Abdessaied, N., Soeken, M., Drechsler, R.: Quantum circuit optimization by Hadamard gate reduction. In: Reversible Computation, pp. 149–162. Springer, New York (2014)
4. Abdessaied, N., Soeken, M., Thomsen, M.K., Drechsler, R.: Upper bounds for reversible circuits based on Young subgroups. Inf. Process. Lett. **114**(6), 282–286 (2014)
5. Abdessaied, N., Soeken, M., Drechsler, R.: Technology mapping for quantum circuits using Boolean functional decomposition. In: Reversible Computation, pp. 149–162. Springer, New York (2015)
6. Abdessaied, N., Soeken, M., Dueck, G.W., Drechsler, R.: Reversible circuit rewriting with simulated annealing. In: International Conference on Very Large Scale Integration, pp. 286–291. IEEE, New York (2015)
7. Abdessaied, N., Amy, M., Soeken, M., Drechsler, R.: Complexity of reversible circuits and their quantum implementations. Theor. Comput. Sci. **618**, 85–106 (2016)
8. Abdessaied, N., Amy, M., Soeken, M., Drechsler, R.: Technology mapping of reversible circuits to Clifford + T quantum circuits. In: International Symposium on Multiple-Valued Logic. IEEE (2016, accepted)
9. Abdessaied, N., Miller, D.M., Soeken, M., Drechsler, R.: Optimization of NCV and Cliffford + T quantum circuits (in preparation)
10. Amy, M., Maslov, D., Mosca, M., Roetteler, M.: A meet-in-the-middle algorithm for fast synthesis of depth-optimal quantum circuits. Trans. CAD Integr. Circuits Syst. **32**(6), 818–830 (2013)
11. Amy, M., Maslov, D., Mosca, M.: Polynomial-time T-depth optimization of Clifford + T circuits via matroid partitioning. Trans. Comput.-Aided Des. Integr. Circuits Syst. **33**(10), 1476–1489 (2014)
12. Arabzadeh, M., Saeedi, M., Zamani, M.S.: Rule-based optimization of reversible circuits. In: Asia and South Pacific Design Automation Conference, pp. 849–854 (2010)
13. Arabzadeh, M., Zamani, M., Sedighi, M., Saeedi, M.: Logical-depth-oriented reversible logic synthesis. In: Proceedings of the International Workshop on Logic and Synthesis (2011)
14. Arabzadeh, M., Saheb Zamani, M., Sedighi, M., Saeedi, M.: Depth-optimized reversible circuit synthesis. Quantum Inf. Process. **12**(4), 1677–1699 (2013)

© Springer International Publishing Switzerland 2016
N. Abdessaied, R. Drechsler, *Reversible and Quantum Circuits*,
DOI 10.1007/978-3-319-31937-7

15. Barenco, A., Bennett, C.H., Cleve, R., DiVinchenzo, D., Margolus, N., Shor, P., Sleator, T., Smolin, J., Weinfurter, H.: Elementary gates for quantum computation. Am. Phys. Soc. **52**, 3457–3467 (1995)

16. Bell, J.S.: Speakable and Unspeakable in Quantum Mechanics: Collected Papers on Quantum Philosophy. Cambridge University Press, Cambridge (2004)

17. Bennett, C.H.: Logical reversibility of computation. IBM J. Res. Dev. **17**(6), 525–532 (1973)

18. Berut, A., Arakelyan, A., Petrosyan, A., Ciliberto, S., Dillenschneider, R., Lutz, E.: Experimental verification of landauer's principle linking information and thermodynamics. Nature **483**, 187–189 (2012)

19. Biere, A., Cimatti, A., Clarke, E., Zhu, Y.: Symbolicv model checking without BDDs. In: Tools and Algorithms for the Construction and Analysis of Systems, vol. 1579, pp. 193–207. Springer, Heidelberg (1999)

20. Bocharov, A., Svore, K.M.: A depth-optimal canonical form for single-qubit quantum circuits (2012). arXiv preprint arXiv:1206.3223

21. Bozzano, M., Bruttomesso, R., Cimatti, A., Junttila, T., van Rossum, P., Schulz, S., Sebastiani, R.: The mathsat 3 system. In: Conference on Automated Deduction, pp. 315–321. Springer, New York (2005)

22. Bravyi, S., Kitaev, A.: Universal quantum computation with ideal Clifford gates and noisy ancillas. Phys. Rev. A **71**, 022316 (2005). doi:10.1103/PhysRevA.71.022316. http://link.aps.org/doi/10.1103/PhysRevA.71.022316

23. Brummayer, R., Biere, A.: Boolector: an efficient SMT solver for bit-vectors and arrays. In: Tools and Algorithms for the Construction and Analysis of Systems, pp. 174–177. Springer, New York (2009)

24. Bruttomesso, R., Cimatti, A., Franzén, A., Griggio, A., Sebastiani, R.: The MathSAT 4 SMT solver. In: Computer Aided Verification, pp. 299–303. Springer, New York (2008)

25. Bryant, R.E.: Graph-based algorithms for Boolean function manipulation. IEEE Trans. Comp. **35**(8), 677–691 (1986)

26. Buhrman, H., Cleve, R., Laurent, M., Linden, N., Schrijver, A., Unger, F.: New limits on fault-tolerant quantum computation. In: Symposium on Foundations of Computer Science, pp. 411–419. IEEE, New York (2006)

27. Chakrabarti, A., Sur-Kolay, S.: Nearest neighbour based synthesis of quantum Boolean circuits. Eng. Lett. **15**, 356–361 (2007)

28. Chuang, I.L., Yamamoto, Y.: A simple quantum computer (1995). arXiv preprint quant-ph/9505011

29. Clarke, E.M., Biere, A., Raimi, R., Zhu, Y.: Bounded model checking using satisfiability solving. Formal Methods Syst. Des. **19**(1), 7–34 (2001)

30. Cook, S.A.: The complexity of theorem-proving procedures. In: ACM Symposium on Theory of Computing, pp. 151–158. ACM, New York (1971)

31. Curtis, H.A.: A New Approach to the Design of Switching Circuits. van Nostrand, Princeton, NJ (1962)

32. Datta, K., Gokhale, A., Sengupta, I., Rahaman, H.: An esop-based reversible circuit synthesis flow using simulated annealing. In: Applied Computation and Security Systems, pp. 131–144. Springer, New York (2015)

33. Datta, K., Sengupta, I., Rahaman, H.: A post-synthesis optimization technique for reversible circuits exploiting negative control lines. Trans. Comput. **64**(4), 1208–1214 (2015)

34. Davio, M., Thayse, A., Deschamps, J.P.: Discrete and switching functions. McGraw-Hill, New York (1978)

35. Davis, M., Logeman, G., Loveland, D.: A machine program for theorem proving. Commun. ACM **5**, 394–397 (1962)

36. Davis, M., Putnam, H.: A computing procedure for quantification theory. J. ACM **7**, 506–521 (1960)

37. de Moura, L.M., Bjørner, N.: Z3: an efficient SMT solver. In: Tools and Algorithms for the Construction and Analysis of Systems, pp. 337–340. Springer, New York (2008)

38. De Vos, A.: Reversible Computing: Fundamentals, Quantum Computing and Applications. Wiley, London (2010)
39. De Vos, A., Van Rentergem, Y.: Young subgroups for reversible computers. Adv. Math. Commun. 2(2), 183–200 (2008)
40. Desoete, B., De Vos, A.: A reversible carry-look-ahead adder using control gates. Integr. VLSI J. 33(1), 89–104 (2002)
41. Deutsch, D., Jozsa, R.: Rapid solution of problems by quantum computation. R. Soc. Lond. Ser. A: Math. Phys. Sci. 439(1907), 553–558 (1992)
42. Devitt, S.J.: Classical control of large-scale quantum computers. In: International Conference Reversible Computation, pp. 26–39 (2014)
43. Dürr, C., Heiligman, M., Høyer, P., Mhalla, M.: Quantum query complexity of some graph problems. In: Automata, Languages and Programming, pp. 481–493. Springer, New York (2004)
44. Dutertre, B., De Moura, L.: The yices SMT solver. Tool paper at http://yices.csl.sri.com/tool-paper.pdf 2, 2 (2006)
45. Eén, N., Sörensson, N.: An extensible SAT solver. In: SAT 2003. LNCS, vol. 2919, pp. 502–518. Springer, New York (2004)
46. Fazel, K., Thornton, M., Rice, J.: Esop-based Toffoli gate cascade generation. In: Pacific Rim Conference on Communications, Computers and Signal Processing, pp. 206–209 (2007)
47. Fowler, A., Devitt, S., Hollenberg, L.: Implementation of shor's algorithm on a linear nearest neighbour qubit array. Quantum Inf. Comput. 4(4), 237–251 (2004)
48. Fowler, A.G., Stephens, A.M., Groszkowski, P.: High-threshold universal quantum computation on the surface code. Phys. Rev. A 80(5), 052312 (2009)
49. Fredkin, F.F., Toffoli, T.: Conservative logic. Int. J. Theor. Phys. 21(3/4), 219–253 (1982)
50. Gaidukov, A.: Algorithm to derive minimum ESOP for 6-variable function. In: International Workshop on Boolean Problems, pp. 141–148 (2002)
51. Giles, B., Selinger, P.: Exact synthesis of multiqubit Clifford + T circuits. Phys. Rev. A 87(3), 032332 (2013)
52. Golubitsky, O., Maslov, D.: A study of optimal 4-bit reversible toffoli circuits and their synthesis. Trans. Comput. 61(9), 1341–1353 (2012)
53. Gosset, D., Kliuchnikov, V., Mosca, M., Russo, V.: An algorithm for the T-count. Quantum Inf. Comput. 14(15–16), 1261–1276 (2014)
54. Große, D., Wille, R., Dueck, G.W., Drechsler, R.: Exact multiple control Toffoli network synthesis with SAT techniques. Trans. Comput.-Aided Des. Integr. Circuits Syst. 28(5), 703–715 (2009)
55. Große, D., Wille, R., Dueck, G.W., Drechsler, R.: Exact synthesis of elementary quantum gate circuits. J. Multiple-Valued Log. Soft Comput. 15(4), 283–300 (2009)
56. Grover, L.K.: A fast quantum mechanical algorithm for database search. In: The Twenty-Eighth Annual ACM Symposium on Theory of Computing, pp. 212–219. ACM, New York (1996)
57. Haedicke, F., Frehse, S., Fey, G., Große, D., Drechsler, R.: metaSMT: focus on your application not on solver integration. In: International Workshop on Design and Implementation of Formal Tools and Systems (2011)
58. Häffner, H., Hänsel, W., Roos, C.F., Benhelm, J., al kar, D.C., Chwalla, M., Körber, T., Rapol, U.D., Riebe, M., Schmidt, P.O., Becher, C., Gühne, O., Dür, W., Blatt, R.: Scalable multiparticle entanglement of trapped ions. Nature 438, 643–646 (2005)
59. Hirata, Y., Nakanishi, M., Yamashita, S., Nakashima, Y.: An efficient method to convert arbitrary quantum circuits to ones on a linear nearest neighbor architecture. In: International Conference on Quantum, Nano and Micro Technologies, pp. 26–33. IEEE, New York (2009)
60. Hirayama, T., Nishitani, Y.: Exact minimization of AND-EXOR expressions of practical benchmark functions. J. Circuits Syst. Comput. 18(3), 465–486 (2009)
61. Jones, N.C.: Logic synthesis for fault-tolerant quantum computers (2013). arXiv preprint arXiv:1310.7290
62. Kane, B.: A silicon-based nuclear spin quantum computer. Nature 393, 133–137 (1998)

63. Khan, M.H.A.: Cost reduction in nearest neighbour based synthesis of quantum Boolean circuits. Eng. Lett. **16**, 1–5 (2008)
64. Kirkpatrick, S., Gelatt, C.D., Vecchi, M.P.: Optimization by simulated annealing. Science **220**(4598), 671–680 (1983)
65. Kliuchnikov, V., Maslov, D., Mosca, M.: Fast and efficient exact synthesis of single-qubit unitaries generated by Clifford and T gates. Quantum Inf. Comput. **13**(7–8), 607–630 (2013)
66. Knill, E., Laflamme, R., Milburn, G.J.: A scheme for efficient quantum computation with linear optics. Nature **409**(1), 46–52 (2001)
67. Knuth, D.E.: The Art of Computer Programming, vol. 4A. Addison-Wesley, Upper Saddle River, NJ (2011)
68. Laforest, M., Simon, D., Boileau, J.C., Baugh, J., Ditty, M., Laflamme, R.: Using error correction to determine the noise model. Phys. Rev. A **75**, 133–137 (2007)
69. Landauer, R.: Irreversibility and heat generation in the computing process. IBM J. Res. Dev. **5**(3), 183–191 (1961)
70. Larrabee, T.: Test pattern generation using Boolean satisfiability. Trans. Comput.-Aided Des. Integr. Circuits Syst. **11**(1), 4–15 (1992)
71. Lindgren, P., Drechsler, R., Becker, B.: Improved minimization methods of pseudo kronecker expressions for multiple output functions. In: International Symposium on Circuits and Systems, vol. 6, pp. 187–190 (1998)
72. Marques-Silva, J., Sakallah, K.: GRASP: A search algorithm for propositional satisfiability. Trans. Comput. **48**(5), 506–521 (1999)
73. Maslov, D.: Reversible logic synthesis benchmarks page. Available at http://webhome.cs. uvic.ca~dmaslov/. Last accessed Jan 2011
74. Maslov, D.: Reversible logic synthesis. Ph.D. thesis, University of New Brunswick (2003)
75. Maslov, D., Dueck, G.: Improved quantum cost for n-bit toffoli gates. Electron. Lett. **39**, 1790 (2003)
76. Maslov, D., Dueck, G.W.: Reversible cascades with minimal garbage. Trans. Comput.-Aided Des. Integr. Circuits Syst. **23**(11), 1497–1509 (2004)
77. Maslov, D., Miller, D.M.: Comparison of the cost metrics through investigation of the relation between optimal NCV and optimal NCT three-qubit reversible circuits. IET Comput. Digit. Tech. **1**(2), 98–104 (2007)
78. Maslov, D., Dueck, G., Miller, D.: Simplification of toffoli networks via templates. In: Symposium on Integrated Circuits and Systems Design, pp. 53–58 (2003)
79. Maslov, D., Miller, D.M., Dueck, G.W.: Fredkin/Toffoli templates for reversible logic synthesis. In: International Conference on Computer Aided Design, pp. 256–261 (2003)
80. Maslov, D., Dueck, G.W., Miller, D.M.: Toffoli network synthesis with templates. Trans. Comput.-Aided Des. Integr. Circuits Syst. **24**(6), 807–817 (2005)
81. Maslov, D., Young, C., Dueck, G.W., Miller, D.M.: Quantum circuit simplification using templates. In: Design Automation and Test in Europe, pp. 1208–1213 (2005)
82. Maslov, D., Dueck, G.W., Miller, D.M.: Techniques for the synthesis of reversible toffoli networks. Trans. Des. Autom. Electron. Syst. **12**(4), 42 (2007)
83. Maslov, D., Dueck, G., Miller, D., Negrevergne, C.: Quantum circuit simplification and level compaction. Trans. Comput.-Aided Des. Integr. Circuits Syst. **27**(3), 436–444 (2008)
84. Meter, R.V., Oskin, M.: Architectural implications of quantum computing technologies. ACM J. Emerg. Technol. Comput. Syst. **2**(1), 31–63 (2006)
85. Miller, D.M., Dueck, G.W.: Spectral techniques for reversible logic synthesis. In: International Symposium on Representations and Methodology of Future Computing Technology, pp. 56–62 (2003)
86. Miller, D.M., Sasanian, Z.: Lowering the quantum gate cost of reversible circuits. In: International Midwest Symposium on Circuits and Systems, pp. 260–263. IEEE, New York (2010)
87. Miller, D., Thornton, M.: QMDD: a decision diagram structure for reversible and quantum circuits. In: International Symposium on Multiple-Valued Logic, pp. 30–30 (2006)

88. Miller, D.M., Maslov, D., Dueck, G.W.: A transformation based algorithm for reversible logic synthesis. In: Design Automation Conference, pp. 318–323 (2003)
89. Miller, D.M., Wille, R., Dueck, G.W.: Synthesizing reversible circuits for irreversible functions. In: Euromicro Conference on Digital System Design, Architectures, Methods and Tools, pp. 749–756. IEEE, New York (2009)
90. Miller, D.M., Wille, R., Drechsler, R.: Reducing reversible circuit cost by adding lines. In: International Symposium on Multiple-Valued Logic, pp. 217–222. IEEE, New York (2010)
91. Miller, D.M., Wille, R., Sasanian, Z.: Elementary quantum gate realizations for multiple-control Toffoli gates. In: International Symposium on Multiple-Valued Logic, pp. 217–222. IEEE, New York (2011)
92. Miller, D.M., Soeken, M., Drechsler, R.: Mapping NCV circuits to optimized Clifford $+ T$ circuits. In: Reversible Computation, pp. 163–175. Springer, New York (2014)
93. Mishchenko, A., Perkowski, M.: Fast heuristic minimization of exclusive-sums-of-products. In: International Workshop on Applications of the Reed-Muller Expansion in Circuit Design, pp. 242–250 (2001)
94. Mishchenko, A., Steinbach, B., Perkowski, M.A.: An algorithm for bi-decomposition of logic functions. In: Design Automation Conference, pp. 103–108 (2001)
95. Moskewicz, M., Madigan, C., Zhao, Y., Zhang, L., Malik, S.: Chaff: engineering an efficient SAT solver. In: Design Automation Conference, pp. 530–535 (2001)
96. Mottonen, M., Vartiainen, J.J.: Decompositions of general quantum gates. In: Trends in Quantum Computing Research, chap. 7. NOVA Publishers, New York (2006). http://www.citebase.org/abstract?id=oai:arXiv.org:quant-ph/0504100
97. Nakahara, M., Ohmi, T.: Quantum computing: from linear algebra to physical realizations. CRC Press, West Palm Beach, FL (2008)
98. Nielsen, M., Chuang, I.: Quantum Computation and Quantum Information. Cambridge University Press, Cambridge (2000)
99. Niemann, P., Wille, R., Drechsler, R.: On the Q in QMDDs: efficient representation of quantum functionality in the QMDD data-structure. In: Reversible Computation, pp. 125–140. Springer, New York (2013)
100. Patra, P., Fussell, D.S.: On efficient adiabatic design of MOS circuits. In: Information, Physics, and Computation. Citeseer (1996)
101. Peres, A.: Reversible logic and quantum computers. Phys. Rev. A (32), 3266–3276 (1985)
102. Prasad, M.R., Biere, A., Gupta, A.: A survey of recent advances in SAT-based formal verification. Int. J. Softw. Tools Technol. Transf. 7(2), 156–173 (2005)
103. Rahman, M.M., Dueck, G.W.: An algorithm to find quantum templates. In: Congress on Evolutionary Computation, pp. 1–7. IEEE, New York (2012)
104. Rahman, M.M., Dueck, G.W.: Properties of quantum templates. In: Reversible Computation, pp. 125–137. Springer, New York (2013)
105. Rahman, M.M., Dueck, G.W., Horton, J.: Exact template matching using graphs. Tech. rep., Technical Report TR13–224, Faculty of Computer Science, University of New Brunswick (2013)
106. Rahman, M.M., Dueck, G.W., Horton, J.D.: An algorithm for quantum template matching. ACM J. Emerg. Technol. Comput. Syst. 11(3), 31 (2014)
107. Saeedi, M., Markov, I.: Synthesis and optimization of reversible circuits-a survey. ACM Comput. Surv. 45(2), 21 (2013)
108. Saeedi, M., Zamani, M.S., Sedighi, M., Sasanian, Z.: Reversible circuit synthesis using a cycle-based approach. J. Emerg. Technol. 6(4), 13 (2010)
109. Sarkar, M., Ghosal, P., Mohanty, S.P.: Reversible circuit synthesis using ACO and SA based quine-McCluskey method. In: Midwest Symposium on Circuits and Systems, pp. 416–419 (2013)
110. Sasanian, Z.: Technology mapping and optimization for reversible and quantum. Ph.D. thesis, University of Victoria (2012)

111. Sasanian, Z., Miller, D.M.: NCV realization of MCT gates with mixed controls. In: Pacific Rim Conference on Communications, Computers and Signal Processing, pp. 567–571. IEEE, New York (2011)

112. Sasanian, Z., Miller, D.M.: Reversible and quantum circuit optimization: a functional approach. In: Reversible Computation, pp. 112–124. Springer, New York (2013)

113. Sasao, T.: AND-EXOR expressions and their optimization. In: Sasao, T. (ed.) Logic Synthesis and Optimization, pp. 287–312. Kluwer Academic Publisher, Dordecht (1993)

114. Sasao, T.: An exact minimization of AND-EXOR expressions using BDDs. In: International Workshop on Applications of the Reed-Muller Expansion in Circuit Design, pp. 91–98 (1993)

115. Sasao, T.: An exact minimization of AND-EXOR expressions using reduced covering functions. In: International Symposium on the Synthesis and Simulation Meeting and International Interchange, pp. 374–383 (1993)

116. Sasao, T.: EXMIN2: a simplification algorithm for exclusive-OR-sum-of-products expressions for multiple-valued-input two-valued-output functions. IEEE Trans. Comput. Aided Des. **12**(5), 621–632 (1993)

117. Sasao, T., Matsuura, M.: DECOMPOS: an integrated system for functional decomposition. In: International Workshop on Logic Synthesis, pp. 471–477 (1998)

118. Scott, N., Dueck, G., Maslov, D.: Improving template matching for minimizing reversible toffoli cascades. In: International Reed-Muller Workshop (2005)

119. Scott, N., Dueck, G., Maslov, D.: Improving template matching for minimizing reversible toffoli cascades. In: International Symposium on Representations and Methodology of Future Computing Technologies (2005)

120. Selinger, P.: Quantum circuits of T-depth one. Phys. Rev. A **87**(4), 042302 (2013)

121. Shannon, C.E.: A symbolic analysis of relay and switching circuits. Trans. Am. Inst. Electr. Eng. **57**(38–80), 713–723 (1938)

122. Shende, V.V., Prasad, A.K., Markov, I.L., Hayes, J.P.: Synthesis of reversible logic circuits. Trans. Comput.-Aided Des. Integr. Circuits Syst. **22**(6), 710–722 (2003)

123. Shi, J., Fey, G., Drechsler, R., Glowatz, A., Hapke, F., Schloffel, J.: PASSAT: efficient sat-based test pattern generation for industrial circuits. In: Computer Society Annual Symposium on VLSI, pp. 212–217. IEEE, New York (2005)

124. Shiou-An, W., Chin-Yung, L., Sy-Yen, K., et al.: An XQDD-based verification method for quantum circuits. IEICE Trans. Fundam. Electron. Commun. Comput. Sci. **91**(2), 584–594 (2008)

125. Shor, P.W.: Algorithms for quantum computation: discrete logarithms and factoring. Found. Comput. Sci. 124–134 (1994)

126. Smith, A., Veneris, A., Fahim Ali, M., Viglas, A.: Fault diagnosis and logic debugging using Boolean satisfiability. Trans. Comput.-Aided Des. Integr. Circuits Syst. **24**(10), 1606–1621 (2005)

127. Soeken, M., Thomsen, M.K.: White dots do matter: rewriting reversible logic circuits. In: Reversible Computation, pp. 196–208. Springer, New York (2013)

128. Soeken, M., Wille, R., Dueck, G., Drechsler, R.: Window optimization of reversible and quantum circuits. In: International Symposium on Design and Diagnostics of Electronic Circuits and Systems, pp. 341–345 (2010)

129. Soeken, M., Frehse, S., Wille, R., Drechsler, R.: Revkit: a toolkit for reversible circuit design. J. Multiple-Valued Log. Soft Comput. **18**(1) (2012). RevKit is available at http://www.revkit.org

130. Soeken, M., Sasanian, Z., Wille, R., Miller, D.M., Drechsler, R.: Optimizing the mapping of reversible circuits to four-valued quantum gate circuits. In: International Symposium on Multiple-Valued Logic, pp. 173–178. IEEE, New York (2012)

131. Soeken, M., Wille, R., Hilken, C., Przigoda, N., Drechsler, R.: Synthesis of reversible circuits with minimal lines for large functions. In: Asia and South Pacific Design Automation Conference, pp. 85–92. IEEE, New York (2012)

132. Soeken, M., Wille, R., Otterstedt, C., Drechsler, R.: A synthesis flow for sequential reversible circuits. In: International Symposium on Multiple-Valued Logic, pp. 299–304. IEEE, New York (2012)
133. Soeken, M., Miller, D.M., Drechsler, R.: Quantum circuits employing roots of the Pauli matrices. Phys. Rev. A **88**, 042322 (2013)
134. Soeken, M., Abdessaied, N., Drechsler, R.: A framework for reversible circuit complexity. In: International Workshop on Boolean Problems, pp. 123–128 (2014)
135. Soeken, M., Tague, L., Dueck, G.W., Drechsler, R.: Ancilla-free synthesis of large reversible functions using binary decision diagrams. J. Symb. Comput. **73**, 1–26 (2016)
136. Soeken, M., Wille, R., Keszocze, O., Miller, D.M., Drechsler, R.: Embedding of large Boolean functions for reversible logic. ACM J. Emerg. Technol. Comput. Syst. **12**(4), 41 (2015)
137. Stergiou, S., Papakonstantinou, G.: Exact minimization of esop expressions with less than eight product terms. J. Circuits Syst. Comput. **13**(01), 1–15 (2004)
138. Szyprowski, M., Kerntopf, P.: Low quantum cost realization of generalized Peres and Toffoli gates with multiple-control signals. In: Conference on Nanotechnology, pp. 802–807. IEEE, New York (2013)
139. Toffoli, T.: Reversible computing. In: de Bakker, W., van Leeuwen, J. (eds.) Automata, Languages and Programming, p. 632. Springer, New York (1980). Technical Memo MIT/LCS/TM-151, MIT Lab. for Comput. Sci.
140. Van Rentergem, Y., De Vos, A., Storme, L.: Implementing an arbitrary reversible logic gate. J. Phys. A Math. Gen. **38**(16), 3555–3577 (2005)
141. Vandersypen, L.M.K., Steffen, M., Breyta, G., Yannoni, C.S., Sherwood, M.H., Chuang, I.L.: Experimental realization of Shor's quantum factoring algorithm using nuclear magnetic resonance. Nature **414**, 883 (2001)
142. Vemuri, N., Kalla, P., Tessier, R.: BDD-based logic synthesis for LUT-based FPGAs. ACM Trans. Des. Autom. Electr. Syst. **7**(4), 501–525 (2002)
143. Viamontes, G.F., Markov, I.L., Hayes, J.P.: Quantum Circuit Simulation. Springer, Dordrecht/Heidelberg/London/New York (2009)
144. Weinstein, Y.S.: Non-fault tolerant t-gates for the [7,1,3] quantum error correction code. Am. Phys. Soc. **87**(3), 032320, 6 (2013).
145. Wille, R., Drechsler, R.: BDD-based synthesis of reversible logic for large functions. In: Design Automation Conference, pp. 270–275. ACM, New York (2009)
146. Wille, R., Große, D.: Fast exact Toffoli network synthesis of reversible logic. In: International Conference on Computer Aided Design, pp. 60–64 (2007)
147. Wille, R., Große, D., Teuber, L., Dueck, G.W., Drechsler, R.: RevLib: an online resource for reversible functions and reversible circuits. In: International Symposium on Multiple-Valued Logic, pp. 220–225. IEEE, New York (2008). RevLib is available at http://www.revlib.org
148. Wille, R., Soeken, M., Drechsler, R.: Reducing the number of lines in reversible circuits. In: Design Automation Conference, pp. 647–652. IEEE, New York (2010)
149. Wille, R., Soeken, M., Przigoda, N., Drechsler, R.: Exact synthesis of Toffoli gate circuits with negative control lines. In: International Symposium on Multiple-Valued Logic, pp. 69–74. IEEE, New York (2012)
150. Wille, R., Soeken, M., Otterstedt, C., Drechsler, R.: Improving the mapping of reversible circuits to quantum circuits using multiple target lines. In: Asia and South Pacific Design Automation Conference, pp. 145–150 (2013)
151. Wille, R., Lye, A., Drechsler, R.: Considering nearest neighbor constraints of quantum circuits at the reversible circuit level. Quantum Inf. Process. **13**(2), 185–199 (2014)
152. Wille, R., Soeken, M., Miller, D.M., Drechsler, R.: Trading off circuit lines and gate costs in the synthesis of reversible logic. Integr. VLSI J. **47**(2), 284–294 (2014)
153. Yamashita, S., Markov, I.L.: Fast equivalence-checking for quantum circuits. In: IEEE/ACM International Symposium on Nanoscale Architectures, pp. 23–28. IEEE, New York (2010)
154. Yamashita, S., Minato, S.i., Miller, D.M.: Synthesis of semi-classical quantum circuits. J. Multiple-Valued Log. Soft Comput. **18**(1) (2012)

155. Yanushkevich, S.N., Miller, D.M., Shmerko, V.P., Stankovic, R.S.: Decision Diagram Techniques for Micro-and Nanoelectronic Design Handbook. CRC Press, West Palm Beach, FL (2005)
156. Zhang, J., Sinha, S., Mishchenko, A., Brayton, R., Chrzanowska-Jeske, M.: Simulation and satisfiability in logic synthesis. In: Proceedings of Workshop on Logic and Synthesis, pp. 161–168 (2005)
157. Zhirnov, V.V., Cavin, R.K., Hutchby, J.A., Bourianoff, G.I.: Limits to binary logic switch scaling-a gedanken model. IEEE **91**(11), 1934–1939 (2003)

Printed in the United States
By Bookmasters